TEST YOUR SCIENCE IQ

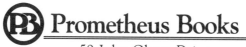

59 John Glenn Drive
Amherst, New York 14228-2197

Published 2000 by Prometheus Books

Test Your Science IQ. Copyright © 2000 by the Estate of Charles J. Cazeau. All rights reserved. No part of this publication may be reproduced, stored in a retrieval system, or transmitted in any form or by any means, digital, electronic, mechanical, photocopying, recording, or otherwise, conveyed via the Internet or a Web site, without prior written permission of the publisher, except in the case of brief quotations embodied in critical articles and reviews.

Inquiries should be addressed to
Prometheus Books
59 John Glenn Drive
Amherst, New York 14228–2197
VOICE: 716–691–0133, ext. 207
FAX: 716–564–2711
WWW.PROMETHEUSBOOKS.COM

04 03 02 01 00   5 4 3 2 1

Library of Congress Cataloging-in-Publication Data

Cazeau, Charles J.
  Test your science IQ / Charles J. Cazeau
     p.   cm.
  Includes index.
  ISBN 1–57392–851–8 (pbk. : alk. paper)
  1. Science–Miscellanea. I. Title.

Q173 .C39 2000
500—dc21   00–057580
                                                                        CIP

Printed in the United States of America on acid-free paper

To

Scott, Annie, and Jeff

CONTENTS

Preface   9

Acknowledgments   11

PART ONE: OUTER SPACE   13

   1. Beyond the Solar System   15
   2. Within the Solar System   35

PART TWO: THE EARTH   59

   3. The Stuff the Earth Is Made Of   61
   4. A Mobile Earth   87

5. Ice, Landslides, Floods, and Other Annoyances   101
6. A Few Sketches: Physics and Chemistry   121

PART THREE: LIFE ON EARTH 141

7. Genes, Survival, and Extinction   143
8. Animals: Sea, Land, and Air   155
9. The Green Earth: Plants   181
10. Nutrition and Health   193

PART FOUR: THE EMERGENCE OF HUMANITY 215

11. Precivilization: Hunters and Gatherers   217
12. Early Civilizations: Getting Organized   231
13. Belief and Behavior   251

PART FIVE: THE PARANORMAL 273

14. Magic, Witches, and Devils   275
15. Monsters   295
16. Ghosts, Haunted Places, and Life Beyond   305
17. Prophecy, Superstition, and the Mind   317

Epilogue   353

Index   355

PREFACE

The purpose of this book is not only to entertain, but also to inform the reader. It is a wide-ranging ramble through the world of science, with special emphasis on those questions that most provoke the curiosity of the individual (we wouldn't mind if scientists buy it too). No single volume can hope to address all of human scientific knowledge, but this one volume covers a fair amount of ground.

This book is divided into five major parts. Part One considers the universe beyond the solar system, and then our solar system itself. Galaxies, stars, and planets are surveyed concerning their origin and nature, and the data obtained from space probes.

Part Two deals with Earth itself, a major subject. Separate sections examine the materials that make up Earth. We will discover the interplay of forces that create earthquakes, volcanoes, and erosion. In short, we will show Earth as the great natural laboratory it is, where physics, chemistry, geology, and biology dynamically combine.

Life itself, both animals and plants, are considered in Part Three, which describes its mysterious origin, evolution, and diversity. A special section is devoted to human health and nutrition. Following this is Part Four, reviewing the findings of anthropology and archaeology in tracing human development during the past two million years, both prior to civilization and in forming ancient civilizations through today.

Finally, in Part Five, the great sphere of the unknown is explored. We examine magic, witches, the devil, ghosts, prophecy, haunted houses, and ouija boards. What we do not know and hope to discover is laid before the reader.

The question-and-answer format lends itself well to an informal style, hopefully direct and understandable, with an avoidance of abstruse terminology that only serves to confound rather than enlighten the reader. These questions came from readers of a variety of newspaper columns I wrote over a period of fifteen years. My answers have been updated and revised for the enjoyment and, I hope, entertainment of the reader.

Charles J. Cazeau

ACKNOWLEDGMENTS

I wish to thank all those who helped in the preparation of this work. I especially thank the staff of Prometheus Books, particularly Executive Editor Linda Greenspan Regan, who monitored all aspects of the book from its inception. I would like to thank Nancy England of Oak Ridge, Tennessee, for her insight, and Peter Avery, geologist, Buffalo, New York, for his encouragement and expertise. Finally, I would like to extend special thanks to my daughter, Suzanne C. Cazeau, whose aid in final manuscript preparation and indexing was invaluable.

PART 1

OUTER SPACE

1

BEYOND THE SOLAR SYSTEM

Q. *Is the big bang theory for the origin of the universe still valid?*

A. Yes it is, although scientists still ponder this subject and admit that we have no sure answers. The stars are moving outward from each other at incredible speeds as though all matter in the universe was at one time packed together and then, in one short burst, began expanding outward in response to a gigantic cosmic explosion—the big bang. This is believed to have occurred between 10 to 15 billion years ago. Recent support bolstering this theory came from measurement of microwave background radiation in space that is presumed to be remnant radiation left over from the big bang.

Conversely, some scientists believe that we live in a pulsating universe. At some point in the far distant future, the stars will collapse back together in what is known as the *big crunch,* and then "explode" again. This cycle could then repeat itself. Scientists have abandoned an earlier theory known as the *steady state* theory. It held that matter was continually being created in outer space and expanding outward.

Q. *Just how many stars are there in a galaxy, and how many galaxies are there?*

A. The Milky Way galaxy in which we are located is estimated to contain 100 billion stars, a number that defies the imagination. And the Milky Way is by no means the largest galaxy. The galaxies, aptly termed *island universes* for their relative density compared to the rest of space, run into the billions. A definite number is hard to come by because as technology improves and we probe further into deep space, we find more galaxies. Interestingly, we were not even certain galaxies existed as late as the 1920s. Through the telescopes of the time, the galaxies appeared only as fuzzy and distant stars. We now know that many galaxies are disclike pinwheels, such as the Milky Way, while others tend to be globular or of irregular shape.

Galaxies are not distributed randomly through space, but gather in clusters. Stars also gather in clusters. Our Milky Way is one of a cluster of about thirty-five galaxies. Even so, the distances from one galaxy to another can be 10 million light-years. The distance from one galactic cluster to another may be ten times or more of that distance.

Q. *With so many billions of stars out there, how do they form? They must have had a beginning.*

A. Star formation begins within a dense cloud, or nebula, consisting chiefly of hydrogen and dust particles. This interstellar matter concentrates under the influence of gravity and eventually becomes increasingly hot to the point of thermonuclear reactions. Hydrogen is the fuel, converting into helium by fusion. At some stage, equilibrium is achieved between the outward pressure of radiation and the downward force of gravity, a condition known as the *main sequence*. The star may remain in this state of balance for many millions of years. When the hydrogen is mostly used up, collapse takes place and at the core, helium can be converted into heavier elements such as carbon and oxygen.

Depending upon its size, mass, or other stellar attributes, the star in its end-state may become a very dense white dwarf slowly cooling through the brown and black dwarf stages to become a burned-out cinder. Otherwise, in more massive stars, the conversion to heavier elements at the core may involve the production of iron, and a large explosion, or nova, may result. In some cases, a black hole might become the final fate of a star. We might note that blue stars are the hottest, white or yellow stars are intermediate in temperature, and reddish stars are the coolest. Whatever the color, they are all exceedingly hot by Earth standards.

Q. *Even with all the stars out there, the vast distances suggest that most of the universe is empty space, right?*

A. You would be a little off the mark with that conclusion, but not entirely. We see in the heavens matter giving off light. There is just as much, maybe more, interstellar matter which is called *dark*. We know that it is there from the gravitational effects it has on the surrounding luminous matter. It is a paradox that an average cubic mile of space contains a mere three or four milligrams of hydrogen and dust. A small pinchful of matter in an essentially perfect vacuum. Yet this is sufficient to contribute significantly to the amount of matter necessary to generate stars by the billions. Powerful telescopes such as the Hubble Space Telescope show luminous galaxies and nebula with large dark patches silhouetted against them. This is another way dark interstellar matter betrays its presence. Still, there is a lot of empty space in proportion to matter. Isaac Asimov once likened the observable universe to a building 20 miles long, 20 miles wide, and 20 miles high containing a single grain of sand.

Q. *Just what is a black hole, and if they are so black, how do we know they are there?*

A. The thought that these objects might exist in outer space was first mentioned about two hundred years ago by Pierre-Simon de Laplace, and later revived in 1916. The existence of black holes is logical. A star uses its nuclear fuel and in that process the outward pressure counterbalances the gravitational force of the mass of the matter that makes up the star. When the nuclear fuel is all used up, gravity takes over and the former star collapses into a small ball of incredibly dense matter. For example, to make Earth a black hole, the substance of Earth would have to be compressed into a ball about one-half inch in diameter! Its force of gravity is so great that not even light can escape, so the black hole is invisible.

As to how we know of their existence, we are aware of many things that exist but are nonetheless invisible to us, like the air. We become aware of them by indirect means; for example, the movement of tree limbs in a breeze tells us that air is real. Likewise, we can detect black holes by their gravitational pull on nearby stars. There are several candidates for black holes, including Centaurus X-3, but the best one found so far is called Cygnus X-1. Some scientists believe that the center of the Milky Way galaxy is the site of a black hole.

Beyond the Solar System

Q. *Television sets have been named after them, but I still don't know what a quasar is—do you?*

A. Yes and no. They were first discovered during the 1960s by radio telescope, but their nature was unknown. They were named *quasars* as shorthand for *quasi-stellar radio sources*, and appeared to be incredibly bright objects at a distance from Earth of 10 billion light-years (more than 60 trillion miles away). They also are traveling at unbelievable velocities, such as 177,000 miles per second. Because they are so distant, they may be objects formed early in the history of the universe, and the light we are receiving from them now was traveling through space for several billion years before Earth itself came into existence.

Further study of quasars indicates that they are extremely massive objects with a luminosity one thousand times greater than an average star, and are imbedded within an otherwise ordinary galaxy. They seem to be about the size of our own solar system which is not surprisingly large. Much yet remains to be understood in the study of quasars.

Q. *Why are radio telescopes treated as more important than optical telescopes?*

A. Perhaps because they are more recent and also have detection capability of celestial activity beyond the reach of optical telescopes equipped with lenses. The latter telescope has been around since about 1608, whereas the radio telescope did not come into wide use until the mid-twentieth century. Nearly all astral objects display radio-wave emissions, from heat or electron activity. Evaluation of received radio waves can tell us a great deal about the nature of stars and galaxies. Once radio waves were found emitting from the Sun and the center of the Milky Way, they were also found emitting from

100,000 other sources, most of them in deep space beyond the Milky Way.

Radio waves are of a longer wavelength than those in the visible light spectrum. They travel at the speed of light and are of a great variety. The discovery of a general background of the shorter-length microwaves in space offered further proof to scientists that these microwaves were a kind of afterglow left over from the big bang. Radio waves also leave Earth and travel outward in space. It may be that thousands of years from now a dweller on a distant planet might listen in on *The Lone Ranger* or one of our favorite soap operas.

Q. *Is there really something called* **antimatter,** *or is it just a science fiction device?*

A. Antimatter exists and is real. The matter we are familiar with consists of such basic atomic particles as electrons, protons, and neutrons. Antimatter is just the same except it carries an opposite positive electric charge. These antimatter particles are known as *positrons, antiprotons,* and *antineutrons.* When matter and antimatter come together, there is an instantaneous annihilation of the particles and an intense release of energy in the form of, for example, gamma rays. Antimatter has been created in physicists' laboratories using particle accelerators. They have even made an antimatter hydrogen atom. What is frustrating for scientists studying antimatter is that the life span of the particles is so short it is hard to study them.

It is believed that many of the explosions seen in distant galaxies are resulting from matter and antimatter coming together. Even in the Milky Way there is a large plume of suspected antimatter rising from the galaxy's center. It is theorized that at the time of the big bang, there were about equal amounts of matter and antimatter produced, but most of the antimatter has been eliminated. Today, the occurrence of antimatter is on the order of only one part to 107 parts of our matter. Yet it is still being created anew, perhaps in supernova explosions where particles are sufficiently accelerated by the blast to bombard other particles to produce antimatter. We can speculate that there may be entire galaxies consisting solely of antimatter. Thus, there could be worlds harboring antimatter life. It would be dangerous for us to shake hands with them.

Q. *What is a white dwarf star?*

A. It is an older, lower-mass star that has passed through the stage of helium-hydrogen consumption and undergoes collapse. What we are looking at is the core of the star after the outer envelope of matter has blown away during a Red Giant stage. This core, thought to be composed mostly of carbon and oxygen, may have the mass of an average star, but be only the size of Earth. This means the core has an extreme density estimated to be one million times that of water. A degenerative electron gas exerting its pressure from the core counterbalances the gravitational force of this star. Such a dwarf star may remain in that condition, slowly losing its residual heat, until it becomes a black dwarf, and in that sense, a dead star.

However, if the mass is great enough, the electrons at the core may join with protons to form neutrons and become a neutron star, an even denser star than the white dwarf. If heat buildup is sufficient, nuclear fusion may cause the formation of heavier elements, including iron, and a catastrophic explosion might occur. Otherwise, the neutron star may contract enough to form a black hole. Some neutron stars that are rotating and emitting regular pulses are referred to as *pulsars*.

Q. *If the stars we see at night are moving as fast as astronomers say they are, why are they always in the same place?*

A. They are moving at breathtaking speeds, as we know from the red shift. Some of the more distant stars are racing through space at many thousands of miles per second. Their apparent lack of movement is due to the vast distances that separate us. Apparent motion is a function of distance and the relationship between the observer and the moving object. Anything moving away from the observer lacks any great motion. The stars are moving away from us. An example from our common experience is a jet airplane taking off and receding from us. The farther away it gets, the less movement you see. Indeed, in many cases, the aircraft seems to be standing still in the sky. Yet we know the plane is traveling at speeds of 400 to 500 miles per hour. Thus, stars many billions of miles away also appear motionless. But given enough time, they will shift position. The present north star, Polaris, has not always been the pole star, and over thousands of years, all stars will be in new positions.

Q. *You mentioned the* **red shift.** *What exactly is this?*

A. The red shift and the Doppler effect can be mentioned in the same breath as they are closely related. They both describe the behavior of waves from an object in motion to a stationary observer. In the case of the Doppler effect, we are referring to sound waves. With the red shift, we are talking about light waves. Christian Johann Doppler was an Austrian physicist and professor at the University of Vienna. In 1842, he noted that sound waves change pitch in moving toward and away from an observer. A simple example within our common experience would be a train whistle approaching an observer with the pitch or tone rising higher, and then in passing, the pitch recedes to a lower level. Similarly, light waves arriving from a star moving away from us (as they all do) will show a displacement toward the red end of the visible spectrum, which represents the longer waves. If instead all the stars and galaxies were rushing toward Earth, we would speak of the *violet shift* in the spectrum. Recognition of the red shift and its use in astronomy was the foundation for the concept of the expanding universe. Edwin Hubble was the first astronomer in about 1929 to put the red shift to practical use. Hubble developed a mathematical constant demonstrating that the farther away from us the galaxies are, the faster they are traveling.

Q. *I know they named a telescope after the astronomer Hubble. Why was this honor bestowed on him?*

A. Edwin Hubble died in 1953, but he is considered one of the leading astronomers of the twentieth century. He graduated from the University of Chicago, majoring in mathematics and astronomy, but did not immediately go into astronomy as a career. He became a lawyer for a while but did not like it and returned to Chicago for further training in astronomy. He served in the armed forces during World War I. In 1922, he became associated with Mount Wilson Observatory, and made a monumental discovery that many of the fuzzy objects thought to be stellar accumulations within the Milky Way were in actuality other galaxies lying well beyond the Milky Way. Right away this enlarged our vision of the size of the universe. Hubble spent some time classifying and sorting these other galaxies based on size and shape. In 1929, he showed that the red shift disclosed the velocity and distance of galaxies, and that they were moving outward and away from us. Thus, the idea of an expanding universe became a foundation stone of cosmology.

The telescope named in his honor is a wondrous tool, an orbiting reflecting telescope situated 370 miles above Earth and so able to see clearly through space without the hindering and obscuring atmosphere to get in the way. It has two cameras and two spectrographs and is able to detect objects fifty times fainter than heretofore. The photographs taken by this telescope of the planets and other objects in the cosmos are spectacular.

Q. *I've read that if you traveled in a spaceship at the speed of light, you would age much more slowly compared to people back on Earth. Can you explain?*

A. This concept is part of Einstein's theory of relativity which has revolutionized our ideas about space and time. We would first caution that, according to Einstein, nothing could travel at or exceed the speed of light, because in so doing you would achieve infinite mass, an impossibility. The fastest celestial objects in the universe are the outermost galaxies and quasars, moving at 90 percent of the speed of light (close to 168,000 miles per second).

Einstein said that the faster you move, especially as you approach the speed of light, the more time slows down relative to a stationary observer. This applies to biological processes as well as mechanical objects such as clocks. Thus, a person aboard a spaceship, moving close to the speed of light, will experience a slowing of heartbeat and respiration. An eight-hour sleep would seem of several days' duration to the stationary observer. In theory, you could return to Earth and be younger than your grandchildren.

The flow of time is not fixed and unchanging as we all thought because we are used to Earth rates of travel. Einstein was proved correct as early as 1936 at Bell Laboratories when radioactive hydrogen atoms were accelerated to high velocities and compared to hydrogen atoms at rest. The frequency of vibration (a sort of clock) of the moving atoms was reduced as Einstein predicted. When asked, Einstein said, "Is it any more strange to assume that moving clocks slow down than to assume that they don't?" Whenever Einstein's theories have been put to the test, they have been shown to be correct.

Q. *Are planets rare freaks of nature or are they common? If there are not many of them, the possibility of life elsewhere would be small.*

A. Your point is well taken. How common are planets? No one knows for sure because at interstellar distances, they are tiny, elusive objects. Yet we are here and our sun does have a family of planets. Are we therefore freaks? At the turn of the twentieth century, scientists were inclined to believe so. It was held that our sun came into a near collision with a passing star. This was conjectured to have caused gigantic gaseous eruptions from the Sun, eruptions that fell into orbit around the Sun and cooled down to form the planets. An accident, in other words. This idea has been abandoned. It is now believed that planetary systems are a normal part of stellar evolution, and if that is the case, then there are billions of planets circling stars, and the possibility of other life is magnified. An exception is the double-star system, where two stars orbit each other. The gravitational interplay in this situation is too complex to favor planet generation. At least half of all the stars we see in the sky are double-star systems. Even so, if we take the dim view that only one set of planets is produced per galaxy, and of these only one out of a thousand can support life, that still leaves millions of planets capable of harboring some life-forms.

Q. *I know that some scientists think there is life on other planets. Isn't this a sort of probabilities guessing game?*

A. Not entirely. Extragalactic planets have already been found. If we are talking about our kind of life, that life is based upon the elements carbon, oxygen, hydrogen, and nitrogen with a little phosphorus and sulfur. These combine to form such vital things as proteins, water, fat, and carbohydrates needed to construct life on Earth whether it is human, cockroach, or oak tree. Analysis of light from distant stars indicates that these elements are present throughout the universe. At this point, the numbers game comes in. If there were only a few planets, the chance that any of them would be earthlike and possess these important elements would be pretty slim. But the trillions of stars we know exist suggest that there must be also at least billions of planets of varying composition. As the number of these planets increases, so too does the possibility that there are planets by the millions with the right ingredients for life to start, given enough time. Life exists on Earth. Why should we be so unique?

Q. *How many planets have now been found beyond the solar system?*

A. At least twenty-nine would qualify as planets. However, some may be simply the burned-out remains of the smaller companion in a double star. What we do know about them is that they tend to be larger than our own planets—several times the size of Jupiter—and pursue strongly elliptical rather than circular orbits around the parent star. There is little chance any would harbor life with such eccentric orbits, as the planet would be too hot and then too cold during a single circuit of the parent star. These suspected planets were discovered during the 1990s and set off an aggressive search by astronomers for other stars with planets. The closest star with an orbiting planet is about eight light-years away. None of them have revealed a family of planets like we have in the solar system. Nonetheless, planets exist besides our own, and radio telescopes in the years ahead may find some surprises.

Q. *Is there a possibility that life could exist on a planet so cold that any water present would take the form of ice?*

A. Scientists have thought about this. We regard a life-form as an energy system that requires some kind of liquid to carry out the chemical reactions that make life possible. Here on Earth, life is essentially protein-in-water. Water is the vital liquid.

On a planet where water is always in a solid form (ice), there could not be earth-type life. However, bear in mind that other substances such as ammonia and methane, which are gases on Earth but would be liquids on a cold planet, could be the basis for some kind of life on a cold planet. With ammonia, this could be an ammonia-in-protein kind of life at –30°F; with methane, it could be a form of life involving fatty substances in methane at –300°F. We can only guess what form such life would take, and whether or not it would be intelligent. One thing is certain: such life would regard our Earth as intolerably hot. We may never know if such life exists or not.

Q. *Could life occur on planets so hot that there are gases but no liquids which are necessary for life?*

A. We cannot conceive of life being possible on entirely gaseous planets. Jupiter in our own solar system is an example. We would have to say, though, that on certain hot planets (by our standards), the idea of life might be entertained if sulfur and silicon were in liquid form. These elements become liquid at temperatures ranging from 235°F to 800°F and above. Both are, of course, solid at Earth temperatures. Yet both are *joiners* like carbon, the basis for Earth life. A *joiner* is an element that can enter into an astounding number of combinations with other elements to form a variety of substances, including substances allowing the possibility of life. What this means is that on some remote, scorched planet there may be life-forms, including intelligent life, that might look at our world and declare it to be too cold to harbor life "as they know it." One is reminded of the story of alien space voyagers making a quick scan of Earth and announcing it to be devoid of life. "Too much oxygen," they reasoned. Oddly enough, oxygen is highly corrosive and might justifiably be imagined as unfriendly to the generation of life. Yet without it, we would perish.

Q. *I have heard that regular radio signals are constantly bombarding us from outer space. Could these be attempts to communicate by an alien intelligence?*

A. There is a very good chance that they are not. Outer space is full of radio noise generated by the activities of astral bodies carrying out their "life" cycles. If you wish to receive an instant message of these radio transmissions, look at the "snow" on your television set or listen to the static of a radio broadcast from a distant station. Much of this is from outer space. Yet it is true many radio transmissions are very regular or have some kind of pattern. Many stars have pulses, such as pulsars, that occur like clockwork, to the second. We know these are not the doings of intelligent life. Here on Earth, we can observe many cyclic events. Old Faithful geyser, for example, erupts at fairly regular intervals and without any intelligent beings working the valves. This is not to dismiss the possibility of life elsewhere in the universe. However, planetary systems with life would be so incredibly far apart that communication even by radio would be a tall order. Although radio waves travel at the speed of light, it would take years, even centuries, for messages to traverse back and forth.

Q. *Why are scientists deliberately using radio waves to try and contact aliens in outer space? Could this invite danger?*

A. You have probably heard of the SETI program (search for extraterrestrial intelligence). It should be pointed out that

SETI is chiefly a listening-in program to discover if any radio waves incoming from space bespeak an intelligent origin. With the myriad of radio signals over millions of channels it is, as has been said often, like searching for a needle in a haystack. One approach has been to focus the big dishes on narrow frequencies of less than one Hertz because these would be least likely to be of natural origin. Very large dishes such as the one in Puerto Rico (1,000 feet across) can focus very well and at great range with high sensitivity but cover only a small part of the sky. In contrast, smaller dishes can cover a larger part of the sky but with much lower sensitivity. A problem is with analyzing results. What kind of signal would another civilization send and could we understand it? Some SETI analysts opine that signals already received and noted might have been signals from another civilization, but not recognized as such.

Without much fanfare, the SETI program has mushroomed into a worldwide effort to include amateurs from 220 different countries and involving more than 200,000 people. A program known as SETI@home has enlisted the help of interested individuals equipped simply with a personal computer and/or a satellite dish. Using an installed software program, your computer, while in screensaver mode, can be receiving radio signals, recording, and returning them for analysis. Whether such an effort will pay off is a moot question. Nonetheless, it is possible some day it will be announced that a grandmother in New Jersey has received a message from an advanced outer-space civilization.

As to the danger of this, the popular media seem determined to portray most aliens as a danger and menace to us earthlings. It seems more likely that they would be, as an advanced civilization, benevolent and helpful. Indeed, they may wish to avoid us in looking at our own history of war, aggression, bloodletting, and mayhem.

Q. *If an alien UFO were to land on Earth, what would its creatures probably look like?*

A. It is unfortunate in our literature that the term *UFO* has come to be synonymous with *alien spaceship*. Many forget that the term stands for *unidentified flying object*, with our emphasis on the word *unidentified*. Given, however, that such a UFO was a space vehicle, it would be an unlikely, nearly impossible event. The vast distances, involving trillions of miles in all directions, make such an encounter, even with good navigation, supplies, and a fast ship, highly unlikely. There is also the time factor. It takes light traveling at 186,000 miles per second many years and decades to reach Earth, even from the restricted neighborhood of our own galaxy. Hence, several generations living and reproducing aboard ship may well be required to complete such a trip. It would be a heroic effort, even by a technologically advanced civilization. This brings up another point. If the universe is teeming with life as many scientists believe, then why would a particular civilization single us out as an interesting place to visit when they have a million other planets to choose from?

As to what they would look like, we think extraterrestrial life could be of almost any form imaginable. Think of Earth's diversity of life, such as the whale, maple tree, and ant. These have evolved on a planet under the sway of that common environment. Even more bizarre forms of life might arise on a planet whose environment differs from that of Earth. If we are talking about mobile, intelligent life, some scientists argue that the humanoid form has many advantages. Our major sense organs are "clustered" in one package (the head) elevated above ground level. Evolutionary development might favor this arrangement in coping with environmental situations on other planets as well as our own. Hands with opposable thumbs for grasping and making tools would be another big advantage. While these seem like sensible ideas, it remains only speculation without facts.

Q. *Could there have been other civilizations in the universe that arose and died out long before our own?*

A. We think so. The universe was in existence for 10 billion years before Earth was born. If it took our planet nearly 5 billion years to produce intelligent life, then on that scale there were many planets continuously emerging with the potential to develop intelligent life over at least the 5 to 7 billion years prior to Earth's formation. Once the stage of intelligence and rational thought is reached, a civilization can burst forth with lightning speed. The earliest humans possessed intelligence, and that was only a few million years ago. Only 20,000 years ago, humans lived as primitive hunters and gatherers. By 3500 B.C.E., early civilizations were already established, building cities and pyramids, creating art, and developing science. It was only a couple hundred years ago that we did not yet have electricity, radio, combustion engines, or aircraft. In the twentieth century we developed nuclear power; advanced medicine, television, and computers; launched satellites; and sent spacecraft to the Moon.

The advances surely to be made over the course of the twenty-first century bedazzle the mind. Imagine then another equally intelligent civilization that pursued a similar accelerating path to ours but advanced even one thousand years beyond where we are now. Such civilizations may still exist while others died out for various reasons long before humankind made a spear. Yes, we think it possible other civilizations have appeared and disappeared over eons of time before Earth. Not only should there be civilizations far ahead of us, but also a large number who have yet to reach our level of technology.

Q. *How would you compare astrology and astronomy, both of which study the heavens?*

A. The early astronomers in Mesopotamia and China made accurate observations about the stars and planets, and this was science. But they personified the objects they studied as gods and goddesses and surmised they had a direct influence on human affairs. At that point, science went out the window. In contrast, astronomy looks at the origin, evolution, composition, and motions of stars and planets as a part of the workings of nature and apart from the superstitious fantasies that invaded astrology.

What sets astronomy apart is that, as a science, it is subject to change based on new evidence gathered by observation and measurement. Thus, astronomy is like all the sciences, dynamic and in a state of constant revision as the facts demand. Astronomy demonstrated that Earth is not the center of the universe. It eventually showed us that the universe is unbelievably huge and is expanding at a dizzying pace. On the other hand, the basic tenets of astrology do not change. Mars will always be the god of war, imparting aggressiveness to those under its influence. Cancerians will always be influenced by the security of a home and family. No new data is gathered that might change these views. In that sense, astrology cannot be looked upon as a science, but rather an amusement for some, and a serious matter for the gullible.

Q. *Astronomers seem to say a lot about the beginnings of the universe, but not much about how it may end. What do scientists think may happen?*

A. The answer for the moment seems to lie in the idea of entropy. Entropy measures the amount of unavailable energy in a system. There is both ordered and disordered energy. Ordered energy can do work, disordered energy cannot. For example, a tank of gasoline is ordered energy because it can do work exploding in your car's engine to drive pistons and make the car move. The same gasoline burned up and taken to a new energy form cannot do any further work and is disordered. Our sun has ordered energy, part of which we receive on Earth and use it to do work. The solar energy that radiates away into outer space is lost and that energy becomes disordered.

With increase in disorder, there is an increase in entropy. Scientists believe that the universe is trending in the direction of maximum entropy (disorder). Another way to express this is *equilibrium*, a fundamental concept of science. Equilibrium is a condition of rest or balance between two systems. It seems that throughout nature, we see a striving toward equilibrium. On Earth, a river flows downslope, cutting a valley and eroding (ordered energy). When the land becomes flat, the stream stops and rests. It has achieved a state of equilibrium. When the Sun finally burns up all its thermal energy, becomes a white dwarf and this eventually stops glowing with its residual heat, the Sun has also achieved a state of equilibrium and no more energy is available.

Thus, we can imagine the universe running down, with increasing entropy. Motions slow and cease, stars wink out, new galaxies stop forming, and the entire system of our universe achieves a state of equilibrium with nothing to disturb (ordered energy) its rest. At this point, the universe is ended.

2

WITHIN THE SOLAR SYSTEM

Q. *How did the solar system originate?*

A. There is a symmetry to the solar system that bespeaks a common and penecontemporaneous origin for the Sun and the planets as well as associated meteors, asteroids, and comets. The planets all move around the Sun in the same direction, in elliptical orbits, and in the same plane.

With these observations in mind, we can visualize a dense cloud of stellar matter coming together and compacting under the influence of gravity. As this takes place, the cloud rotates faster, and due to centrifugal force there is a flattening of the cloud into a disclike configuration. Within this disc, the protosun stands at the center, accreting most of the infalling matter, while local swirling eddies of matter develop away from the sun, but within the disc. These smaller eddies will become the planets, and have their own gravitational power to attract matter so that they also grow by accretion and at the same time gain heat.

The new sun is so massive, that it triggers off thermonuclear reactions turning the sun luminous. The eddies, now protoplanets, are not sufficiently massive to produce thermonuclear reactions but can reach partially molten conditions. Radiation from the sun blows away lighter gases from the innermost protoplanets (Mercury, Venus, Earth, and Mars), but the outer planets such as Jupiter and Saturn are able to retain much of their lighter gases. The inner planets form atmospheres by gaseous transfer from the planets' interior to the surface. It is in this way, or close to it, that astronomers and astrophysicists think the solar system was formed about 4.5 billion years ago.

Q. *I know that cosmic rays from outer space bombard Earth all the time. What actually are they and can they be harmful?*

A. Cosmic rays are high-speed particles hitting Earth. The particles are of two kinds: 82 to 85 percent of them are protons and are called *primary rays*, while 12 to 16 percent are helium nuclei resulting from collision and are termed *secondary rays*. They are found concentrated within the Milky Way galaxy and appear to be spawned in the course of supernova explosions, which impart to them a very high speed. Upon entering the solar system, they are redirected by magnetic fields in such a way as to pour toward Earth near the top of the atmosphere from all directions Alternatively, scientists are considering whether their high speed is being induced chiefly by magnetic fields.

In any event, they encounter Earth. The primaries

strike other particles to send a shower of secondaries earthward with such speed that many penetrate two miles into Earth's crust. Because primary rays do not reach down to Earth's surface with any abundance, they are best studied from high-altitude balloons or satellites in orbit. It is believed that cosmic rays are relatively young, being little more than 10 million years old. One side effect of cosmic ray bombardment is the production of subatomic particles such as *positrons*, *muons*, and *pions*. This has given rise to the important area of scientific investigation called *particle physics*.

Cosmic rays will deliver a low-level radiation that gives us well below the maximum amount recommended by the U.S. government, and can be considered harmless at such low levels. One can be exposed to a high enough level of radiation due to a job environment or other source, causing such incidents as the disaster that occurred at Chernobyl in the former Soviet Union. After several years of contamination, soil and vegetation are still affected and some people who were exposed to the radiation are dying of cancer.

Q. *What are sunspots and what effect do they have on humans, if any?*

A. One may view these dark spots on the Sun with some reflective material such as tinted glass. They represent great magnetic fields first observed by Galileo, and are known to be considerably cooler than the surrounding sun material. For that reason, some years ago a popular carbonated orange drink was marketed and appropriately named "Sunspot." During peaks of high sunspot activity as many as three hundred spots may be definable. During times of low spot activity only a few may be discernible. The cycle between high and low sunspot activity is about eleven years. *Solar flares* and *prominences* are a spectacular result of sunspot activity. *Solar flares* are violent discharges of energy that take on many forms, such as X rays, ultraviolet rays, and high-speed electrons and protons. *Prominences* are huge clouds of hydrogen gas above the photosphere, which are buoyed up by magnetic forces that carry them off into outer space. Scientists are uncertain as to the cause of sunspot activity.

We seldom think of the Sun as anything more than our source of heat and light. Yet the Sun and sunspot activity cause interference and fade-outs in radio transmissions, the creation of the ozone layer, and noticeable effects on Earth's climate. Tree rings are proof of this in showing growth patterns that follow an eleven-year cycle. It is suspected that sunspot activity brought about a long cold spell in Europe starting in about 1750, during the Little Ice Age. Huge solar flares were the culprits in the 1989 Quebec blackout that left the entire province and peripheral areas of the United States without electrical power. This event cost millions of dollars. So yes, sunspot activity can have a profound impact on Earth.

Q. *Are the northern lights caused by burning gases in the atmosphere?*

A. It is not thought so. We usually think of these nightly luminous displays as occurring in the high latitudes of the Northern Hemisphere as the *aurora borealis*, but equally, they occur in the Southern Hemisphere and there are known as the *aurora australis*. They take the form of arcs, curtains, and other glowing shapes adorned in greenish or reddish hues, having considerable dramatic impact for the observer. While the origin of these phenomena is not fully understood, it is thought that particles (electrons and protons) originating on the Sun, especially during solar-flare activity, are borne to Earth by solar winds. There, under magnetic-field influence, they pour into the polar areas and collide with oxygen and nitrogen of our atmosphere and cause ionization, which in turn causes the colors. The glowing arcs and curtains represent a sort of electrical discharge. It is rather amazing that during a display, 100 million hydrogen particles are striking each square inch of the atmosphere every second. On rare occasions, these lights are seen in southern Europe and the southern United States. In past times, the aurora was believed by the superstitious to be a warning of impending disaster.

Q. *Who first measured the size of Earth and how was it done?*

A. This goes back to the time of the Greeks. Many of them already suspected that Earth was round if for no other reason than that the Sun and Moon, companions of Earth, were round. Some ancient Greeks may have inferred Earth's sphericity from lunar eclipses, where the shadow of Earth falls on the Moon's surface. It is important to know that Earth is round before calculating its size.

The Greek philosopher and mathematician, Eratosthenes, is said to have been the first to scientifically measure the circumference of Earth. He took the distance from Alexandria (in Egypt) up the Nile to what is now Aswan. This distance information was obtained from drivers of camel caravans. To Eratosthenes, this was a segment of arc along a "great circle." The curvature of the arc was obtained by noting the difference in the angle of the sun's rays falling on Aswan, where they were vertical, and at Alexandria, where they were about 7 degrees. He calculated that the circumference of Earth was about 29,000 miles. This was about 2,000 miles too much, but a truly remarkable scientific achievement. It is interesting that earlier peoples thought Earth bigger than it is, but grossly underestimated its age.

Q. *What are the earliest references to eclipses? They must have been frightening to primitive peoples.*

A. The earliest reference we have comes from Chinese astronomers who noted an eclipse on October 22, 2136 B.C.E. So eclipses have been under observation a very long time and many recorded from early civilizations all over the world, including the Greeks, Babylonians, Romans, and Muslims. Quite often, such eclipses were regarded as omens at the time of the death of a king, a victory in battle, or other important events.

We are familiar, of course, with solar and lunar eclipses. The former involves the intervention of the Moon between Earth and the Sun, and the latter, the passing of the Moon into Earth's shadow as cast by the Sun. Astronomers can predict these with great accuracy. We can also include transits of Mercury and Venus across the Sun's disc, called *occultations,* and the eclipsing that occurs in double-star systems where each star revolves around a common center of gravity. Eclipses provide valuable information to the astronomer. Composition and density of planetary atmospheres, size measurements, speed, and other data can be obtained. The rings of Uranus were first observed when that planet eclipsed a bright star.

Yes, eclipses have frightened many throughout the ages. It was believed the Sun was being eaten up, never to return. The sky darkens into a twilight condition, stars come out, and even nocturnal creatures are roused. Birds go to roost. Early peoples beat drums, shouted, and shot arrows into the air to scare away the presumed monster eating up the Sun. The fact that a total solar eclipse lasts only eight minutes added credence to the notion that their efforts were effective. An eclipse of the Moon took place on April 2, 1493. At that time Christopher Columbus was trying to obtain provisions from natives on Jamaica. Knowing of the coming eclipse, Columbus warned the natives he would do something to the Moon unless they gave him food. After the eclipse, Columbus got what he wanted from the frightened natives, who were awed by his power.

Q. *Do most meteors come from outside the solar system?*

A. Most come from within the solar system and represent the leftover debris following the formation of the Sun and its planets. They are abundant, and thousands enter Earth's atmosphere every day, but being mostly small, they burn up in the atmosphere without reaching the ground. This is evident on a dark clear night to the meteor watcher who spots the event as a momentary flash of light, often leaving a long trail of incandescent flotsam. The heat generated is due to friction with the impeding atmosphere. Astronomers like to refer to the object itself as a *meteoroid* and should a remnant reach Earth, then we speak of that as a *meteorite*.

Many of these objects appear to be detached fragments that become strewn along the orbital path of a comet. When Earth enters such a region of space, meteor showers occur. Some meteors could have originated from the disintegration of larger asteroids. There are two major kinds of meteorites, the stony meteorites and the iron-nickel meteorites. Sometimes when these land on Earth they are still hot on the outside but the coldness of the interior soon shows itself as the meteorite accumulates frost on its exterior.

Q. *Could life have begun on Earth because of a meteorite crashing into Earth? There is carbon in some meteorites.*

A. There is carbon in some meteoroids (the iron-nickel variety), but it takes more than carbon for life to form. You also need hydrogen and oxygen to form amino acid and protein. There was no necessity to deliver carbon from outer space since it was already present here in abundance as well as throughout the solar system. The idea that life on Earth could have come from a source in outer space is known as the *cosmozoic* theory and has been held in low regard by scientists for some time. But let us imagine that it happened. The meteorite would have had to survive a fiery trip through the atmosphere and land in the ocean in shallow water. We say this because all the geological evidence shows life began in the sea in a shallow marine environment. A small number of molecule-sized "seeds" would then somehow have to be released from the rock to face an unknown and probably hostile environment. The odds favoring such a development would be astronomical (no pun intended). Scientists have not been able to create life in the laboratory although there have been provocative steps taken in this direction. The origin of life is still a mystery.

Q. *Was the big meteorite that crashed in Siberia at the turn of the century a spaceship? Radioactivity was found indicating atomic power.*

A. It is uncertain if it was actually a meteorite. It may have been a small asteroid or a cometary fragment. The event occurred on the morning of June 30, 1908, in central Siberia near the Tunguska River in a heavily forested area of pine trees and few inhabitants. There were eyewitnesses—some fur traders—who claim it was a bright light with a trajectory of about 30 degrees which blew up with deafening explosions above the ground (there was no crater) and caused high, hot winds and a trembling of the ground as in an earthquake. Indeed, seismographs in Europe recorded the event.

In 1927, a team of Russian scientists visited the area and found that the blast had felled forty-five thousand trees lying in a radial pattern outward from the epicenter like spokes on a wheel. The devastated area was about 800 square miles. It is estimated that the object may have weighed up to one million tons, was traveling at about 62,000 miles per hour, and released the equivalent of 10 to 15 megatons of TNT upon exploding. The meteorite, if that is what it was, vaporized completely. One can imagine the destruction if this event had occurred in a major population area. It may have been the greatest calamity in human history. To conceive of this huge object as a spaceship is quite fantastic. Radioactivity in the area was not considered at an unusual level. Because no crater was formed, it is calculated that the object exploded while at least a few miles above the earth. It is certainly a lesson in the power of nature's fury.

Q. *If Earth is moving so fast through space, how is it we do not feel this motion?*

A. Because we are moving right along with it at the same speed or 66,700 miles per hour, or 18 miles per second. And it is a smooth ride. It is similar to traveling on a jet aircraft at 600 miles per hour and feeling little sense of motion. A fly buzzing around in the cabin of a jet should have even less sense of motion, which is about what we are relative to Earth. It is even more disconcerting when we realize that this is not the only motion we earthlings are subjected to. After all, Earth is spinning on its axis at 1,000 miles per hour and the Sun is carrying the entire solar system along through space while being at the same time involved in the general rotating galactic movement. And of course the galaxy is hurtling outward in the general expansion of the universe. It is certainly enough to make one feel dizzy, yet we are not.

Q. *There was supposed to be a time back in the 1980s when the planets came into alignment and dire things were predicted to occur. Did anything important happen that I missed?*

A. No. During the summer of 1982 the planets Venus, Earth, Mars, Jupiter, and Saturn came into alignment with the

Sun. Seers, psychics, and astrologers came forth, proclaiming great events both good and bad. Neither happened. According to the predictors, persons born during this time would have unusual gifts. Such alignments have taken place many times before and the people born at that time were not particularly distinguished from those born at other times. A cataclysm of earthquakes was supposed to take place because of the combined gravitational pull of all the planets. The planets combined would exert hardly an iota of gravitational force to be noticed in some vivid manner because they represent less than 1 percent of the mass of the solar system. It would be like assuming an elephant would falter in its progress if you threw a cream puff at its legs.

Finally, close analysis shows that contrary to popular conception, the planets were not lined up as perfectly as soldiers in a row. They were imperfectly arrayed within a quadrant of the sky, hardly a strict alignment. It is characteristic that the seers and psychics were strangely quiet in the aftermath of this "alignment." If something unusual had coincidentally occurred, such as a major earthquake, then they would still be talking loudly about it.

Q. *Did Galileo invent the telescope? I know he discovered the moons of Jupiter with a telescope he made personally.*

A. No, he did not. The invention of the telescope was an accident. In 1608, a Dutch eyeglass-maker named Hans Lippershey or one of his assistants, happened to hold up two large lenses while looking out the window and was surprised to see that the view was much closer. Lippershey assembled the lenses in a tube, and the first telescope was made, being of 3-power. Galileo heard of this while in Venice and made his own 32-power telescope, with which he discovered the moons of Jupiter. He observed the phases of Venus, showing that the planets moved around the Sun rather than Earth. The telescope also allowed him to observe the millions of stars in the Milky Way galaxy.

When these discoveries were made public, Galileo was immediately attacked by ecclesiastical authorities because it was contrary to biblical canon. Galileo was a pioneer in mathematics, mechanics, and astronomy, and widely respected during his own lifetime. Had it not been for this and a personal friendship with the pope, he might have been burned at the stake. As it was, he was put under house arrest for the remaining eight years of his life. While such a thing could not happen today, we still see much antiscience in attempts to ban evolution from biology textbooks and the foisting of creationism on students when it has not one scintilla of scientific merit. As one scientist said, "Galileo may be dead, but his persecutors are alive and well."

Q. *Is it possible the Moon started out to be an independent planet but was captured by the Earth's gravitational field and became our satellite?*

A. It is an interesting idea that scientists have considered. There are two main models for the origin of the Moon, the Binary Accretion Model and the Giant Impact Model. The Binary Accretion Model suggests that the Moon and Earth were formed separately at the same time from the same dust particles. The Moon formed close enough to Earth so that it was held in Earth's orbit by gravity. The Giant Impact Model suggests that the young Earth was grazed by a Mars-sized body, which was heavy in iron, spewing forth debris from Earth's mantle, and the resulting debris condensed to form the Moon.

Most scientists today believe that the Moon formed at the same time as Earth, had a similar accretion history, and was about in the same relative position to Earth that it now occupies. Rock samples brought back from the Moon by astronauts seem to verify this. The growth patterns on the skeletons of ancient corals 400 million years old indicate a tidal influence suggesting the Moon was in place at least that long ago. However, their mutual origin goes back further than that to the early days of the solar system when planets were still accreting and growing, gathering space debris unto themselves. Earth, being the larger body, attracted the heavier metallic elements, which is shown by Earth's large metallic core of iron and nickel. On the other hand, the Moon grew by sweeping up the lighter, stony materials still orbiting the heavier Earth. The moon today is a much lighter, less dense body than Earth, suggesting that it had only the lighter material left to incorporate.

Q. What are the large dark spots that make up the eyes of "the man in the Moon"?

A. Early in the Moon's history, during accretionary bombardment, some large asteroid-sized bodies struck its surface with such force that they opened gaps that penetrated to the interior. At that time, the Moon's interior was still molten with lava material. This lava upwelled to the surface, creating "lakes" of the dark lava known as *basalt*. This basalt is quite familiar to us here on Earth as it is the primary material that built the Hawaiian Islands and is still doing so to this day. A characteristic of basalt is its low viscosity, causing it to spread out almost like water during eruptions. It did this on the Moon, then cooled and solidified. Early astronomers called these areas *mare* (pronounced MAR-ay) which means *seas* in Latin, as many thought this was what they were.

Astronauts have landed on the Moon in the mare areas. They found no water but brought samples of mare material back to Earth. These lunar rocks are similar to terrestrial basalt with one major difference. The minerals in the lunar rocks are as fresh looking as when they were formed, because without water and atmosphere there is no weathering on the Moon. Earth basalt shows much alteration of original minerals due to chemical weathering.

Q. *The craters of the Moon seem perfectly circular. How is it that the meteors that made them always descend vertically to the Moon's surface?*

A. You are assuming, not illogically, that circular craters can only be made by vertical impact. Laboratory experiments have been carried out using gas guns and simulated lunar soil to see how these craters might have formed. Results show that if the trajectory of a meteoroid is greater than 15 degrees, a circular crater is always formed. However, if you study photos of the lunar surface carefully, you can see elongated craters or "skip marks" where meteoroids landed at angles of less than 15 degrees. There are also rays of brighter material showing splatter. You will also see that not all craters are fresh and sharp. Many are softened at the edges and rather faded looking. These are older, degraded craters. Their appearance is caused by micrometeorite bombardment, a sort of sandblasting effect that erodes the craters and represents the only erosive activity on the Moon. You can also see that some craters are volcanic in origin and that there are lava flows associated with them.

Q. *When I see all the millions spent on the space program, I ask if this money could not be better spent on Earth? This is especially true of the shuttle program, bringing astronauts back and forth.*

A. We think it is worth the money if for no other reason than it advances human knowledge. It is also a good investment in terms of return on the dollar. The shuttle is actually a relatively cheap way to explore space because the shuttle is reusable and malfunctions of orbiting satellites can be repaired without writing them off. The technology that has been developed and is ongoing is of benefit to everybody. From space exploration, weather and storms can be better predicted and save lives by early warning. Vegetation patterns can aid in the area of agriculture. Satellites can help locate valuable mineral deposits, identify the best fishing grounds in the ocean, and keep an eye on air- and water-pollution trends. Experiments in space are advancing medicine and promoting the development of new products useful to us. Perhaps some day, the space shuttle will be used to assemble components for a manned spaceship that will be launched for interplanetary exploration and even interstellar missions.

Q. *Could the asteroid belt be the remains of an exploded planet?*

A. Most of the asteroids in the solar system are concentrated in this *belt* area which lies between the orbits of Mars and Jupiter. The popularity of this theory of an exploded planet may derive in part from its use as a theme in science-fiction stories. Alternatively, there is the theory that the belt was formed when two extra-large asteroids collided, blowing both of them to smithereens. Neither scenario is seriously considered by astronomers.

Scientists believe that the asteroid belt represents debris from the early solar system that never coagulated into a planet. Despite the fact that, if you count the smaller fragments as well as the large ones, the total mass is less than our own moon. This relatively small mass of material came under the gravitational influence of our largest planet, Jupiter, which kept perturbing the asteroidal orbit and prevented coagulation into a planet.

Q. *How large are the asteroids in the asteroid belt? Could any of them have a chance of hitting Earth?*

A. The largest is Ceres, discovered in 1801, and on the order of 600 miles in diameter. There are about sixteen whose diameters exceed 145 miles. These larger asteroids tend to be spherical in shape like the planets, but the smaller asteroids assume a variety of irregular shapes. The bulk of the asteroids (93 percent) are stony, rich in silicate minerals while only about 6 percent are metallic, consisting of nickel and iron. There are thousands of medium-sized asteroids, and if you count the smaller fragments, there are millions of them.

The orbits of some of these larger asteroids do cross Earth's orbit and open a remote possibility of encounter. Many of the smaller asteroids, as meteoroids, do strike Earth but most burn up in the atmosphere or land somewhere with little or no damage, such as the ocean. An exception is the object that struck in Arizona, forming the large Barringer Crater. It is believed this came from the asteroid belt. In general, a meteoroid striking the surface of any planet will blast out a crater about four times its own diameter.

If we were imagining one of the larger asteroids hitting Earth, it would be a disaster of unparalleled magnitude. If the impact area were the middle of the United States, a crater as wide as 1,500 miles would be created, with probable upsurgings of molten lava and dust clouds girdling Earth. It is doubtful that humanity could survive such a cataclysm. An even worse scenario is the same asteroid landing in the ocean and spawning giant tidal waves that would sweep into coastal areas around the world. With most people living along coastlines, the deaths of millions of people would be certain. Yet, in perspective, such an event might occur only once in a million years. In the case of smaller but still deadly asteroids, disasters may happen once in 10,000 years. The disturbing part of such a statistic is that we never know when something might happen. It could happen next month or next year.

Q. *What is the difference between an asteroid and a comet?*

A. Asteroids are mostly made up of rocky material, are irregular in shape, and generally pursue orbital paths within the solar system. There are hundreds of thousands of them, if not millions. They are the leftover debris from the early development of the solar system. In this respect, most of them are as old as the planets. In contrast, comets are mostly ice, frozen gas, and dust particles, aptly described as "dirty snowballs." The long gaseous tail of the comet is its hallmark. Many comets also follow highly eccentric orbits that take them out of the solar system proper only to return periodically after hundreds of years. Comets are fewer in number than asteroids, but new ones can form. Only about nine hundred comets have been recognized. Comets are short-lived because much of the ice and gas are vaporized by the Sun's heat which leaves only a black carbonaceous nucleus. Few comets last for more than one million years. Some scientists believe comets predate the birth of the solar system. Both asteroids and comets are of varying mass, which is difficult to measure.

Q. *I notice that Mercury has many craters like the Moon. Are they very much alike?*

A. The two bodies are packed with impact craters and they both go through phases as seen from Earth. They are of comparable size and are extremely hot. The surfaces of Mercury and the Moon are subject to micrometeorite bombardment (the sandblasting effect) despite the fact that Mercury has a very thin atmosphere of helium, sodium, and oxygen. This has led to crater degradation of varying degrees in both places and produced a thin layer of dust and fine debris. The surfaces are very old and have remained basically unchanged for millions of years. Other than these general similarities, there are important differences, some of them revealed by the *Mariner 10* probe of 1974–75.

Mercury has a density much like Earth while the moon is considerably less dense. The innermost planet revolves around the Sun in eighty-eight days; the Moon revolves around a common gravitational center with Earth. Mercury has a weak magnetic field, suggesting a metallic core, which may be still partially molten. There are prominent escarpments exceeding a mile in height due to faulting, which are not present on the Moon. The view from the surface of Mercury, looking at the sky, would be quite different. The sky would appear darker, as it is on the Moon, but the Sun would blaze away at 2.5 times larger in apparent size than seen from the Moon. Surprisingly enough, there seems to be some water on Mercury in ice caps shielded from the heat in places not reached by the heat of the Sun. Temperatures range from 100 to 700 Kelvin (about –280°F to 800°F). It would not be a pleasant place to live.

Q. *Venus has a thick atmosphere that hides the surface. What must the surface look like and why does it have a thick atmosphere?*

A. Being the brightest planet in the sky, and known to the earliest civilizations, it is strange that it was also the most mysterious despite being the planet closest to Earth. As you note, the thick, cloudy atmosphere completely obscured any chance of a glimpse of the actual surface. Even the invention of good optical telescopes failed to add very much to our knowledge of the surface. It was, however, discovered that the dense atmosphere was composed mostly of carbon dioxide. Many imagined it to be a hot, steamy planet with jungles and even dinosaurs or other primitive life. This was quite far from the truth.

In the 1960s and 1970s, the Soviets landed a space vehicle on the Venusian surface and the pictures returned revealed a hot, dry, and rocky planet. Further space missions by both the Soviets and Americans, more than twenty in all, opened the door to a wealth of new information and images. This was particularly true of the U.S. *Magellan* spacecraft, which went into orbit around Venus in August 1990 and returned details of the surface using radar, which was able to penetrate the clouds. Now, more than 90 percent of the surface has been mapped.

Venus is much less cratered than Mars or the Moon. Fewer than one thousand craters have been named. The Venusian topography is varied, with high plateaus, valleys, plains, ridges, and scarps. There are numerous volcanoes and lava fields, but there is no water. Thus, the landscape has been shaped chiefly by vulcanism, faulting, and other tectonic activity. This may explain the rich carbon-dioxide atmosphere that resulted from volcanic gases transferred to the surface. Active volcanoes here on Earth also spew carbon dioxide into the atmosphere.

Q. *Wasn't there life, even a great civilization, on Mars? I saw a picture of the face on Mars and I heard that the two satellites are artificial.*

A. The face on Mars turned out not to be a face at all. When the photo was taken from space, the shadows cast from the Sun gave it the appearance of a face. On a later pass, the area was photographed again from a different angle with few shadows, and the object turned out to be nothing more than a jumble of rocks. Besides that, the face had resembled George Washington more than a Martian god.

The two satellites orbiting Mars (Deimos and

Phobos) are certainly not objects launched from the surface of Mars. Even the smaller of the two is nine miles across, while the larger (Phobos) is more than 16 miles at its widest point. This would be a heavy liftoff load. Both satellites are of irregular shape and composed of rock and ice. They are covered with a layer of dust. They have been struck by numerous meteoroids to form craters. These highly unlikely artificial satellites are regarded by astronomers as asteroids escaped from the asteroid belt and captured by Mars. It is clear from orbital speeds and other factors that one day Phobos will crash down on Mars, and its sister satellite will recede farther and farther from the parent planet.

Indeed, life could have started on Mars. Possible microorganisms have been found with the aid of an electron microscope in Martian meteorite material, which strongly resemble similar forms on Earth. Debate continues about whether the microorganisms in the vein of rock originated on Mars or that they infiltrated the rock after landing in the Antarctic 13,000 years ago. The carbonate vein where the microorganisms have been found is dated at 3.9 billion years old, and scientists agree that at that time, Mars had a much warmer and wetter climate. Currently, the climate on Mars is too cold for liquid water (averaging –63°F), but there is frozen water on or near the surface. Pictures taken by the Mars Global Surveyor suggest that water came from underneath the surface and filled the channels. Flash flooding may have caused some of the channels, but we do not know yet if any were carved by prolonged water erosion. It is possible that volcanic activity brought water to the surface, and if there is still volcanic activity, it could mean that there are hot springs close to the surface, thus improving the chances of life on Mars. The topography of Mars has been mapped, finding one area where the surface resembles that of a shoreline, possibly from a long-ago Martian ocean. Mars Polar Lander launched in December 1999 has mysteriously disappeared, its goal to find water ice in the southern pole of the planet. More Mars missions are planned for 2001 and 2003, and more evidence is needed to determine if there is or ever was life on Mars.

Q. *Is that giant volcano on Mars still active?*

A. Not so far as we know but many volcanoes on Earth that were thought to be extinct erupted. The volcano you mentioned is called *Olympus Mons*, and at its height of 17 miles, it is probably the largest volcano in the entire solar system. This is not surprising in view of the fact that many of the surface features of Mars are the result of volcanic and tectonic activity. The planet was able to achieve a molten stage sufficient to form a core, mantle, and crust structurally similar to Earth's except considerably less dense. The gaseous expulsions from the volcanoes formed an atmosphere of carbon dioxide and some nitrogen, oxygen, and rare gases, most of which have been lost.

Particularly intriguing is that early Mars appears to have possessed stream systems. Water may have flowed on the surface and underground. Perhaps during this phase life attempted to develop. However, today all ice is solid due in part to lowered atmospheric pressure. There is ice and carbon dioxide tied up as dry ice in the polar caps and it will probably stay that way. The picture of Mars today is that of a dusty, dry, and rocky planet swept often by wind and dust storms.

Q. *So why is Mars red? And where are the canals astronomers used to report?*

A. Certainly the red coloration is the first thing noted when you look at Mars; the cause was long puzzled over and even today we can't be 100 percent sure. However, the atmosphere of Mars at one time had enough oxygen for it to combine with iron to form iron oxide, or rust, like Earth minerals limonite and hematite. It does not have to be in abundance to cause the color, but serve merely as a coating on grains of other nonred minerals, like paint on a house.

There are no canals on Mars. The entire surface has been mapped in detail by *Mariner* satellites and none have been found. There are, however, linear features such as faults and ridges that early astronomers with less able equipment interpreted as canals. Thus, the idea was hatched that the canals were built by intelligent beings to bring water from the polar caps to an arid civilization. Human vision is not perfect. When viewing something unusual, we tend to link up lines and spots into familiar patterns from within our own experience. We see faces in cloud formations. Faces we see on Mars and the face we see in the Old Man of the Mountain in New England are natural rock formations.

Q. *How many satellites have now been discovered orbiting Jupiter? Could some form of life exist on any of them?*

A. Jupiter is remarkable, not only because it is the largest planet of the solar system, but because it is a solar system in itself, with its sunlike composition (hydrogen-helium) and its retinue of sixteen satellites. There could be more waiting to be found. The innermost group of eight includes the largest ones, first seen by Galileo, and named *Io, Europa, Ganymede*, and *Callisto*. These satellites may have formed like other planets in the solar system, as accretionary condensations in the Jovian nebular cloud. Ganymede and Callisto are as big as the planet Mercury. The outer satellites with their eccentric orbits seem to be captured asteroids.

No life has been detected on the four Galilean satellites although all of them have a cover of water-ice. Callisto and Ganymede are of low relief and show an abundance of craters. Europa, in contrast, has an exceptionally smooth surface and few craters. This suggests to some scientists that Europa is undergoing resurfacing with an outflow of water that quickly forms solid, icy lakes. Io is quite unique in displaying active volcanic eruptions not seen elsewhere in the solar system except on Earth. In 1979, the *Voyager* probe spotted no less than nine eruptions going on, and substantial sulfur and sulfurous compounds that impart a yellow-orange color to the surface.

Q. *What kind of material forms Saturn's famous rings?*

A. When looked at casually, it seems there are only a few major rings. Galileo was the first to set eyes on them and thought they were attached to the surface of Saturn. Christiaan Huygens later said they were detached, which was correct. Later on, seven rings were distinguished, but once the *Voyager* and *Pioneer* space missions reached there, it was found that the rings could be subdivided into as many as one hundred thousand ringlets. These spacecraft also discovered several more satellites around Saturn, bringing the total to eighteen. The ring system extends outward a distance of about 85,000 miles and consists of aggregates of rock fragments, frozen gases, and ice. The fragments range from silt and sand-sized particles up to boulders as big as 35 feet in diameter. In some places, the rings are only 16 feet thick.

Saturn is second only to Jupiter in size and has a similar composition of hydrogen and helium, but somewhat more hydrogen (88 percent). Saturn radiates three times as much heat as it receives from the Sun. Of interest is *Titan*, the largest of the Saturnian satellites. At 3,200 miles in diameter, it is larger than Mercury and has an atmosphere of nitrogen and methane. Some scientists believe Titan may be like Earth was before life began.

Q. *Are the two outer planets, Uranus and Neptune, of similar size and composition?*

A. Uranus is the slightly bigger planet with a diameter of 32,500 miles as opposed to the 30,700-mile diameter of Neptune. In atmospheric composition they are similar—both being composed of hydrogen and helium for the main part—but Neptune's 3 percent methane is different and imbues that planet with a blue tint. Both have several satellites, possibly captured asteroids, Uranus with seventeen and Neptune with eight, and both seem to have an internal heat source. But there are some differences. Uranus is tilted 55 degrees on its axis so that it appears to "roll" along in its orbit about the Sun. There are also nine rings surrounding Uranus. Neptune was predicted to be where it is by mathematical calculation (Urbain–J.J. Leverrier, in 1846) before telescopes were turned to that portion of the sky. This remarkable achievement was based upon distortions in the orbit of Uranus caused by an unseen force.

Q. *The outermost ninth planet, Pluto, has a satellite. How could such a tiny object so far away be detected?*

A. It was again, a case of knowing where to look in the sky after mathematicians said a planet should be there. Clyde W. Tombaugh discovered Pluto in 1930, and its satellite *Charon* was discovered in 1978. Pluto contrasts with the outer planet giants in being a mere 1,430 miles in diameter and Charon is about half that size. Pluto is a cold, rocky world with an ice cap and a thin atmosphere of methane. As is generally known, Pluto sometimes crosses the orbit of Neptune to make Neptune the outermost of the planets. There is a good chance that Pluto is an escaped satellite of Neptune. Some mathematicians still think yet another planet—Planet X—lies beyond the orbit of Pluto, since the irregularities of the two orbits suggest another gravitational force exists. Discoveries within our solar system are not over yet.

Q. *Aside from those who keep predicting an end to the world at the end of every millennium, what would be the dictate of nature in this respect?*

A. For the time being, Earth is alive and well. The existence of Earth is tied closely to the fate of the Sun because the Sun is our source of heat and light. Eventually, however, the Sun will die, and Earth along with it. For example, if the Sun were to suddenly wink out, which wouldn't happen, then Earth would grow colder in an eternal night and life would begin to perish. Earth would become, in due course, a frozen chunk of rock. A few bacteria might survive in a suspended state.

What is more likely to happen is that when the last of the Sun's thermonuclear fuel is consumed, it will enter a nova stage and expand outward beyond Earth's orbit. The oceans would boil and Earth burn to a crisp, and then vaporize—as if it had never been. Our astronomers consider the Sun to be no more than a middle-aged star with several more billions of years of longevity left before a nova stage could occur. Therefore, not to worry.

PART 2

THE
EARTH

3

THE STUFF THE EARTH IS MADE OF

Q. *How old is Earth, and how did it come into existence?*

A. Based upon numerous radiometric dating of rocks from Earth's crust, rock samples brought back from the Moon, and meteorites from surrounding space, scientists have determined the age of Earth to be about 4.5 billion years old. Earth's origin was generally the same as for the rest of the components of the solar system. An immense cloud of hydrogen, dust, and other gases slowly condensed under the influence of gravity, with the center of the cloud becoming the Sun, while smaller eddies outward from the center became the planets. This gigantic cloud was rotating as it condensed, and continued to gain heat. For the Sun the source of this heat was nuclear reaction in the protosun, while the protoplanets gained heat by the unceasing impact of accretion. In other words, planets such as Earth were growing by sweeping up quantities of spatial debris. Perhaps at this stage Earth enlarged itself by at least 25 percent.

Earth itself offers evidence of these long ago events. Our planet, along with the other eight planets, is moving in orbit around the Sun and in the same direction, maintaining the same movement as the original rotating primordial cloud. Earth's interior, down to the center, reveals an onionlike structure of shells, with the heavier elements forming a large core surrounded by a mantle of rocks of lesser density, and a thin crust of the lighter elements. This is exactly what you would expect for a world that passed through at least a partially molten or liquid stage before general solidification. While liquefied, Earth's material was mobile and could move around and settle according to gravitational dictate.

Q. How much of Earth's interior is still liquid?

A. Earth's core can be divided into an inner and an outer core. The inner core is solid, but the outer core is liquid, or at least behaves as a liquid. We cannot of course have a direct view of the core but we can infer conditions there by the behavior of seismic waves generated by earthquakes. When a strong earthquake occurs, it sends out vibrations known as *P waves* and *S waves*. The S waves can travel quickly through solids but are halted if they encounter liquid. In major quakes, when these waves pass through Earth's core, only the P waves arrive at seismic stations on the opposite side of Earth. In short, seismic waves X-ray Earth's insides.

The lava we see pouring out of volcanoes did not travel up from the core, but may have formed from localized "hot spots" due to radioactivity within the crust. Heat to form lava also can originate where continental plates come together, and be fed by heat still being transferred from the interior toward the surface during the process of mountain building. Once lava is formed and is mobile, it is compelled to move toward zones of lesser pressure, like upward to the surface.

Q. *What causes one rock to be different from another?*

A. By definition, a rock is a solid. That is why lava from volcanoes is not considered a rock until it cools and solidifies. As any rock collector knows, there are great and interesting varieties of rock that can be seen in nature. But even so, geologists recognize that no matter where you go on Earth to collect rocks, each can be relegated to one of three broad categories based upon the origin of the rock. These three categories are (1) *igneous* rocks, (2) *sedimentary* rocks, and (3) *metamorphic* rocks.

Igneous rocks are those that have passed through a molten stage before solidifying. The word *igneous* comes from the Latin word *ignis,* meaning "fire." Lava, upon solidifying, becomes an igneous rock. Since the whole Earth has been molten at some time in the past, all rocks have experienced an igneous stage. Granites are igneous rocks. *Sedimentary* rocks are formed as layers of sand, silt clay, or chemical precipitates from any rock that has become exposed and eroded at or near Earth's surface. Shale and limestone are typical sedimentary rocks. *Metamorphic* rocks form from preexisting rock under conditions of high heat and pressure. New platy minerals may appear and structures within the rock take on a parallelism or foliation. The rock may also be described as *imbricate,* like shingles on a roof. Mica schist is an example of a metamorphic rock.

Q. *What is the difference between a rock and a mineral? Or are they the same thing?*

A. No, they are quite different but intimately associated. A mineral is a crystalline solid with a fairly definite chemical formula, and often, each mineral may show distinctive physical properties. Minerals are usually classified by their chemistry into such groups as *silicates, carbonates,* and *sulfides*. All minerals have an outward appearance of color, luster, and hardness. These are the physical properties. Some minerals have special properties that few other minerals have. For example, magnetite is magnetic and halite tastes salty.

You cannot have rocks without minerals because by definition, a rock is an aggregate of minerals. Rocks are named because of their mineral associations. Thus, the igneous rock granite is usually made up of feldspar, quartz, mica, and minor amounts of other minerals, often dark, such as hornblende. Sandstone, a sedimentary rock, may be made up of grains of quartz eroded from granite or other rocks at the earth's surface. Metamorphic rocks may display higher temperature/pressure minerals such as the reddish mineral garnet or a bladed mineral called sillimanite. When some rocks that consist almost entirely of one mineral are found, geologists may fudge a bit on strict definition and refer to it as a *monomineralic* rock. Rocks are also thought of in their definition as integral parts of Earth's crust.

Q. *I was looking at mineral collections in a museum and wondered, why are some crystals of the same mineral bigger or smaller than others?*

A. Hobbyists who grow colorful crystals at home know that these crystals are growing from a solution of water and dissolved ingredients. The same sort of thing happens in nature. Most crystals grow from hot solutions underground. Depending upon what elements are present in the solution, the atoms will arrange themselves in a particular pattern. For example, if silica (oxygen and silicon atoms) is present, six-sided crystals of quartz will form. If the solution cools quickly, small crystals result because the atoms do not have time enough to grow into larger crystals. Slow cooling helps to form large crystals. An important factor is water. The more water in the hot solution, the faster bigger crystals can grow. This is because water makes the atoms much more mobile.

This cooling process has a bearing on valuable ore deposits. In a cooling *magma* (magma is a body of molten rock underground) crystallization takes place. Each mineral that crystallizes out has its own temperature when this happens. Gold, silver, and quartz form at the lower temperatures. So as the magma solidifies and cools, the remaining fluid becomes concentrated in gold, silver, and quartz. This remaining fluid has a great deal of water as well, and an ore body of rock develops containing appreciable amounts of gold and silver. Sometimes huge crystals several feet across of quartz and rarer minerals such as tourmaline and beryl can form. Such late-cooling bodies are called *pegmatites*.

Q. *Why are all snowflakes six-sided?*

A. Not all snow particles are six-sided. Some are pellets or sleety ice. Snow is so familiar to most of us we forget that snow is a mineral. When the opportunity presents itself, water will grow a snow crystal with a six-sided habit and with considerable variety. One oddity of snow is that flakes can fall out of a clear sky when there is not enough water vapor to form clouds, but enough to form on some nucleus such as a speck of dust in the air.

While minerals are usually classified according to their chemical groupings, they can also be classified according to which crystal system they belong to. There are six crystal systems. Snow and quartz both belong to the *hexagonal* system, whose members always show a six-fold symmetry in the crystal state. Another of the six systems is the *isometric* system, in which crystals grow along a threefold axis to form cubes and octagons. Magnetite and halite belong to this system.

Q. *If Earth was once molten then there could be no oceans. Where did all the water come from? And why is it so salty?*

A. Earth first had to form a solid crust with low spots to serve as containers for water. This probably took place within the first 1.5 billion years of Earth's existence. But even after the crust formed there was plenty of molten material stirring around below, giving rise to extensive and prolonged volcanic action bursting through a still very fragile thin crust. In addition to the lava and ash emitted, there were tremendous clouds of gases and water as steam. The steam condensed as it cooled and heavy rains, perhaps at times continuous, started to fill up the future ocean basins. Even today in a modern volcanic eruption, large amounts of water are released into the atmosphere.

At first the water in the growing oceans was fresh. The water, plus the additions from still continuing volcanic activity, began to move in a cycle. We call this cycle the *hydrologic cycle*, where water moves from land to the sea, evaporates, rains down once again on the land, and returns to the sea repeatedly. In returning to the sea, streams and rivers erode and dissolve things like salt. When these same waters evaporate off the ocean surface, they leave the salts behind. In due course, the sea becomes enriched in dissolved salts, and thus saline. This did not happen overnight, but took many millions of years. As volcanic action subsided, the amount of water in the oceans became more or less stable. While to us the oceans seem huge and almost bottomless, this water, at 1,200 feet average depth, is the merest veneer of moisture on a globe 8,000 miles in diameter.

Q. *I was told that the atmosphere was once methane and ammonia that would kill you if you breathed it. How did it change to what it is now?*

A. The envelope of air that surrounds Earth has gone through an evolution but we are not 100 percent certain that in a past age it was completely methane and ammonia. One reason for believing this, however, is that volcanoes expel these noxious gases in eruption. Equally, a case can be made that the primitive atmosphere consisted of nitrogen and carbon dioxide. The history of chemical change in the air is one of great complexity not fully understood. As with the water of the oceans, we turn to volcanic transfer from Earth's interior to explain the presence of an atmosphere. Scientists call this process *degassing*.

In any event, the process of change resulted in an atmosphere of essentially nitrogen with small amounts of carbon dioxide and water vapor, perhaps two billion years ago. With the advent of early plant life in the oceans, oxygen became abundant from the plant life as a by-product of photosynthesis. The earliest plant life, blue-green algae, created "oxygen oases" which were pockets of oxygen surrounding the plants. These became saturated with oxygen and dispersed throughout the waters. Eventually the oceans became saturated with oxygen, and thus began to release it into the atmosphere. Beginning about 600 million years ago there was a remarkable surge and spread of life in the sea and onto the land. The amount of oxygen rose from an estimated 3 percent to its present level of 21 percent. Life had to begin without oxygen, and now most life cannot be sustained without it. Nonetheless, some bacteria do exist today that do not require oxygen to survive. These are referred to as *anaerobic* bacteria.

Q. *I know there are traces of krypton in the atmosphere. What is this gas like?*

A. Yes, krypton occurs in minute traces and is detectable as only one part in 900,000 gas in the atmosphere. It was not discovered until 1898 when some liquid air was almost boiled away and it was found in the residue. It is called a *noble gas*, along with argon, xenon, and neon because these gases do not usually combine with other elements to form compounds. Despite its rarity and the fact that it only liquifies at -242 degrees Fahrenheit, it has uses in industry. It is used in flash lamps for high-speed photography. A major use is in detecting flaws in metal surfaces where a radioactive form of krypton can collect in tiny cracks in the metal and facilitate their identification. We know of no naturally occurring solid krypton, and the small amounts found in the atmosphere probably arrived there from volcanic eruptions and hot springs.

Q. *In talking about science, you hear a lot about using the scientific method. What is so special about the scientific method?*

A. There is nothing mysterious about it. It is a step by step procedure to solve a problem and/or acquire new knowledge. The steps are

1. Identification of a problem or an idea
2. Forming a hypothesis (an explanation)
3. Gathering of data relevant to the hypothesis
4. Testing of the hypothesis against the data
5. Forming a conclusion
6. Verification

Let us take a simple example to show that we all use this method in our daily lives and it is not something that is confined to the laboratories of science. You discover that your wallet (with a lot of money) is missing from the top of your dresser where you last put it. You have a problem

(1, above). Only three people had access to your room, so one of them had to be the culprit (2). Your mother, brother Bill, and cousin Nancy are the three people. On investigation (3) you find your mother was out shopping all day, Bill was in his room reading, and Nancy was doing laundry. Bill also owed a large sum of money to a mutual friend. You check out the kitchen (4) and find new groceries (mother indeed did shopping), there are several piles of cleaned and folded clothes (4) in the laundry room (Nancy did laundry), and you call Bill's friend and find he had paid him some money (4). You conclude that (5) Bill took the money. You confront him and he confesses (6). While this is admittedly an absurd example, it shows how scientists think.

Q. *In scientific thinking, I have heard the expression "Occam's razor." It is not clear to me what this means.*

A. The application of *Occam's razor* can be thought of as an important adjunct to the scientific method and logical reasoning. Sir William of Occam was a fourteenth-century philosopher who stated this as the *principle of parsimony*. What it means is, when confronted with a choice of possible solutions to a problem, the most plausible explanation is the likely answer. For example, a person is missing. Possible explanations include (1) kidnapping or other foul play, or (2) the person was disintegrated by the ray gun of an outer-space alien. Using Occam's razor, we would select (1) as the likely solution. Occam's razor is sometimes described tongue-in-cheek as the Principle of Least Astonishment.

But there is not always total certainty. Suppose while some evidence shows the person could not have been kidnapped or subject to foul play, the government makes an announcement that a fleet of alien spaceships has landed. Now explanation (2) starts to look less astonishing. Similarly, in science, a perfectly plausible theory accepted by many may be rejected reluctantly because it does not fit into new evidence, so a less plausible hypothesis receives greater consideration.

Q. *If life started in the sea, how did it get onto the land? Or did it arise independently on the land?*

A. For millions of years there have been tidal areas along shorelines where land and sea meet. These areas are called *littoral zones*, defined as the area between high and low tides. Here the beach is alternately under water and exposed to the air. It is theorized that certain sea creatures, perhaps fish because of their mobility, became accidentally marooned in this zone at low tide, but were able to survive without water until the next tide came in. With further evolution, these creatures managed to live happily in this mixed environment, and some of them among the fittest and hardiest were able to survive without water for long periods. Eventually, this would give rise to the amphibians (some of whom live today) that were equally at home on land or in water. There are clear evolutionary indications in the fossil record of progression from amphibians to reptiles and mammals.

It is interesting that we carry part of the sea with us, as blood closely resembles seawater. The total amount of water in our bodies is around 70 percent of our body weight. The basic chemical composition of blood is dissolved salts and water. It is doubtful that life arose independently on land. We have no evidence of this. For life to start once is a singular event. For it to happen twice on the same planet is very unlikely. We might speculate on what direction evolution might have taken had there been no moon to cause tides. Would all life still be in the sea and the land barren, and would this aquatic life be as intelligent as we humans are?

Q. *What were the earliest forms of life?*

A. A general rule in evolution is the progression from the simple to the complex, although some seemingly simple creatures such as earthworms are far from simple. The earliest forms were probably bacteria and blue-green algae which have been found in very ancient (3.1 billion years) rocks in the Transvaal of Africa. The first life came from the plant kingdom, with animals following later. But the earliest fossil records are extremely meager. Geologists call this time of scant fossil remains in rocks the Precambrian. In the ensuing Cambrian period, all major phyla were established except the vertebrates. Survivors from that time live in the sea today, as for example, the bottom-dwelling bryozoan and coral groups.

Q. *What must the landscape have been like before plants invaded the land?*

A. The first appearance of any kind of land plant appears to have been around 400 million years ago during the Silurian period. The geologic record shows this in layered sedimentary rocks that preserved these plants as fossils. Land areas must have been bleak, harsh, and uninviting. Also, there would have been no animal life trotting or slithering across the landscape, because plants invaded the land quite a bit before animals. We might imagine such places today, as the Sahara or Atacoma deserts, as equivalent to a Silurian landscape, yet even in those modern places some plants grow and some even thrive. Another aspect of those pre-plant days would have been rapid and deep erosion of the land by running water, because plant cover retards such erosion by holding the soil together with roots. It was the plants both on land and in the sea which gave off oxygen, permitting the invasion of the land by animal life. In time, shore-dwelling amphibians gave rise to reptiles, mammals, and birds, able to range far inland from the sea and become independent of it. Prior to that, all life was confined to the marine environment.

Q. *On a recent fossil-hunting trip I was lucky enough to find a trilobite. Can you tell me something about these interesting fossils?*

A. They first appear in the fossil record of the Cambrian period, 600 million years ago, preserved in limestones and shales. Their flat, oval bodies have a threefold organization consisting of a head, thorax, and tail. This gives them their distinctive appearance. They lived on the muddy sea bottom, able to crawl but not swim. Like other arthropods, trilobites molted, shedding an outer hard skeleton to permit further growth. These molted skeletons are also preserved as fossils as well as tracks and trails where they passed. We might mention here that a *fossil* is any remains of ancient life, direct or indirect. Thus, a footprint, while not the actual remains, still qualifies as a fossil.

Most of these trilobites were only about an inch long but in later periods they grew to as long as two feet. They became extinct at the end of the Paleozoic era, having lived over a span of 400 million years and producing ten thousand species. That is a very successful animal. If you find a trilobite, you know the rock in which it was found is from the Paleozoic era, because they are not found before or after that time. Such fossils are known as *index fossils* and they aid paleontologists in working out the chronology of the rocks in Earth's crust.

Q. *There are some slabby rocks in my back yard that have butterfly fossils in them. How did they get there?*

A. Butterflies and other flying creatures are extremely rare in the fossil record, and for that reason can command a very fancy price. Geologists have established that butterflies and moths are usually no older than 60 million years. In checking a geologic map of the rocks in your area, it seems your butterfly rocks are much older than this. What we think you've got are fossils of a particular order of brachiopod called a *spirifer*, now extinct. These were sea animals that lived on the sea bottom and resembled clams in having two shells that could open and close. The shells, or valves, of these spirifers were often winglike and strikingly resembled butterflies. But this is a superficial resemblance. Spirifers are common as fossils, and their relatives—other brachiopods—live on the seafloor today.

Despite their rarity as fossils, butterflies have been highly successful, generating more than 100,000 different species, and of course they flourish at present. Major reasons for butterfly and moth scarcity as fossils are that they are land/air creatures and have no hard parts such as skeleton or shell. The sea bottom is the most favorable place for fossil preservation. Animals and plants that die on land have only poor chances of becoming a fossil.

Q. *What causes a particular family of plants or animals to become extinct?*

A. When we look at the overall fossil record, especially that covering the past 600 million years, we find that extinction is not some disaster that happens to a few unlucky species. It appears rather to be a part of the natural order of things along with growth and reproduction. Three out of four families of all animals and plants that have ever lived are extinct. One can allude to specific causes such as a cataclysmic encounter with an asteroid, a loss of food supply, the invasion of a predator, or a loss of fertility. Indeed, in individual cases these causes may well apply. However, some geologists point to more general principles that lead groups of the living into extinction.

In evolution, we see a trend over millions of years in which a species becomes better and better adapted to the environmental niche it occupies. A deer, geared for running, develops longer and stronger legs and more enduring lungs through natural selection. A tiger evolves sharper teeth, longer claws, and a keener sense of sight and smell. This is *specialization*. Over the short term this is an advantage to a species. It does one thing well and only that one thing to the exclusion of other traits. Carried to an extreme, we have overspecialization. Should the environmental niche change with relative swiftness, the overspecialized cannot adapt quickly enough. Should plains give way to swamps, for example, the deer's fleetness is lost. A dwindling food supply for the tiger assures extinction when there are no longer prey for sharp claws and teeth to grasp.

Q. *How would it be if extinction never took place? Would we be able to see actual dinosaurs?*

A. Yes, that would be interesting. But keep in mind that extinction is necessary for survival. If that sounds like a contradiction, think again. Imagine today's world as one in which all forms of life from the beginning have survived and live side by side. In swamps, large dinosaurians, lizards, crocodiles, and giant ferns live together while flying pterodactyls share the sky with robins and hawks. Mastodons, giant sloths, elephants, saber-toothed tigers, Neanderthals, bears, wolves, and modern humans rove the land. Every weed that has ever lived competes for space with the corn and carrots of the frustrated farmer. In the sea, plesiosaurs of every description compete with whales of every description for space amid large schools of thousands of species of sharks and bony fishes, both armored and unarmored. The sea bottom is carpeted with every imaginable oyster, clam, and coral. The clutter is beyond belief.

Often, during the course of Earth history, a life-form will appear and give rise to still another life-form with added advantages for survival. Then the original life-form will die out, its task completed. Nature is interested only in survival. It is like opening a can of soup. Once the can (protecting the soup) is opened, it is no longer necessary for the survival of the soup and is discarded. In nature, a fish with legs moves life from sea to land, and is discarded, its job done. Our scenario above is of course pure fantasy. But the point is made that death and removal is as much a part of life as life itself. Nature keeps things tidy. Nature makes room in ecological niches for other species to survive while other preexisting species must go. If it were otherwise, no life could exist on this planet.

Q. *An oil-well worker told me that natural gas and oil come from "bugs." Was he putting me on?*

A. Even if he smiled when he said it, he was telling the truth, in a figurative sense. When we think of bugs, most of us think of such things as flies and mosquitoes, and we doubt seriously that such creatures contributed to our oil supply. However, it is fairly well established that oil and gas have been generated from organic matter, or in other words, once-living creatures. Petroleum geologists theorize that the remains of ancient sea life—particularly little floating forms called *phytoplankton*—became incorporated in bottom sediments, and there were converted into liquid petroleum and gas. Other marine animals may also have contributed.

There is good evidence. Oil and gas are hydrocarbons consisting of the same elements that make up living organisms. The oil also occurs in sediments that formed in environments once teeming with life, such as coral reefs or other shallow marine environments. The successful prospecting and exploration for oil is based on the concept of finding and recognizing ancient rocks that at one time served as shallow marine sites for a prospering biota. Geologists argue about many things, but not this, strange as it may seem.

Q. *Somebody told me that when an oil well is drilled, only about half the oil is recovered. Is this true?*

A. Often, oil companies are fortunate if they get 25 to 35 percent of the amount of oil in place. The oil does not accumulate in "pools" as is commonly believed, but gathers in the small cracks and interstices of the rock. The oil must be forced out of the ground. At first, the difference in pressure between the rock formation and the open boring is enough to move oil toward the well. When this pressure differential declines, the oil will not move. However, it is possible to introduce water into a nearby well under pressure and have the water push out more oil. This is one of a number of methods called *secondary recovery*. Even so, as much as one-third of the oil in place is never recovered.

At one time, the United States was the world's leading producer of oil. In the present, we are dependent upon foreign sources such as the Middle East and Venezuela. The northern slope of Alaska has been a boon to the United States, as has the North Sea for European countries. The United States still has plenty of proved reserves, but the search for new deposits becomes more difficult as the easier-to-find oil is exploited. Oil and gas are nonrenewable resources. Once consumed, they are gone forever. Sooner or later we will run out of oil in the face of rising world demand and will need to rely on alternative fuels such as oil shale and coal, and after that perhaps we will have developed atomic, wind, and solar energy sources to a practical level.

The Stuff the Earth Is Made Of

Q. *Why are coal and oil referred to as fossil fuels?*

A. Oil, natural gas, and coal represent remnants of former life on Earth, thus fulfilling the definition of a fossil. As is generally known, coal beds are the remains of plant life growing under swampy conditions, buried and converted. Conversion passes through a number of stages including peat, cannal coal, bituminous coal, and anthracite. It is not surprising that abundant plant fossils are found associated with coal, except for anthracite, which has undergone the greatest change and the fossils destroyed. The United States has been mining coal, mostly in eastern states, for many years, but still has enormous coal deposits, especially in western states.

Oil and natural gas are harder to imagine as fossil fuels because they are liquid and gas, but geologists do not dispute their organic nature. As in the case of coal, the oil is originally derived from members of the plant kingdom, that is, the phytoplankton of ancient seas. But in this instance, the conversion process results in a liquid rather than a solid. Fossil fuels can be thought of as captured sunlight or *fossil sunlight* that shone upon Earth millions of years ago. This solar energy, converted and stored by coal and oil, is once again released for useful work upon combustion.

Q. *I saw some kind of insect inside a chunk of yellow glass at a curio shop. The owner said that it was an insect that died millions of years ago. Is this possible?*

A. It is likely this was not glass at all, but *amber*. Amber is fossil tree resin or sap exuded from coniferous trees up to 100 million years ago. It occurs in varying shapes, mostly irregular, from droplike to rod-shaped and with a range of colors such as yellow-brown, orange, red-brown, and ochre. Many insects landing upon it while in a sticky condition will be trapped as if on fly paper, and engulfed. Later, the tree sap loses its volatiles (substances that evaporate rapidly), hardens, and becomes amber. The preservation of insects is excellent, revealing the tiniest detail, even the hairs on the leg of a fly. Of course, not all specimens of amber contain an insect. That is the exception. Most in fact are flawed by opaque dust or air bubbles. Clear, transparent, or translucent amber is considered gem quality.

The most famous area for amber is along the shores of the Baltic Sea. More than two thousand species of insects have been found and include flies, gnats, spiders, and mosquitoes that look little different from their modern counterparts, showing that insects have been around a long time without undergoing significant change. These amber deposits are from 40 to 60 million years old. They are searched for along beaches, especially after storms. Known and prized since ancient times, the Romans sent their legions to the Baltic and brought amber back to Rome. The tsars of Russia also collected amber from the Baltic and one palatial room was lined completely in amber and called the Amber Room. The Germans dismantled and carried away this room during World War II and it has not been seen since. It was worth a fortune both historically and intrinsically. Another aspect of amber lore is the possibility that a mosquito in amber from the time of the dinosaurs might have sucked up some dinosaur blood, which is still preserved in the mosquito's body. If so, perhaps it could be extracted and the DNA recovered for cloning. Such a possibility is still only in the imagination.

Q. *I see advertisements selling zircon gems. Aren't these man-made?*

A. You may be referring to cubic zirconia, which is a diamond simulant or artificial crystal resembling a diamond in refraction, dispersion, hardness, and color. Zircons, however, are naturally occurring in many places around the world. They are of varying colors and usually opaque, although sometimes transparent. The gem varieties are sometimes found in river gravel in Australia and New Zealand, but more often in Indonesia and Sri Lanka. These are the principal sources for the metallic element zirconium, which is actually more common on Earth than more familiar metals such as copper. There was not much use for it until the atomic age. Then it was discovered that it was highly useful as a jacket on uranium fuel rods in nuclear reactors.

The most significant source of zircon in the United States is in beach sands along the east Florida coast. The sand-sized particles can be separated from lighter minerals in a centrifuge machine in a modern version of the old gold prospector's panning method. Zircon is not as heavy as gold but it is heavier than quartz along with such minerals as garnet, tourmaline, magnetite, and epidote which are sometimes referred to as *heavy minerals*. There is a good chance that tiny crystals of zircon might be found in the soil of your own backyard, but they would be of little economic value.

Q. *What gives an opal its beautiful play of color?*

A. Oddly enough, it is the imperfections in the stone such as tiny cracks and veinlets that interfere with the light as it passes through, breaking the white light into its spectral colors. Veinlets may contain water or air that contribute to the display of colors. Opals form out of solutions of silica in volcanic areas. It thus has the same elements as the common quartz. Ninety percent of opals come from South Australia. There are numerous varieties of opals with a wide range of colors. The common opal is opaque, with no play of color. Examples include milk opal and honey opal. *Opalescence* refers to color change with orientation. Two subgroups of precious opal are white opal and black opal. The white opal displays flashes of color, while the more rare black opal displays fire with a dark body color. Another type of opal is the fire opal, named for its color, which is transparent but not opalescent.

Although the Romans valued the opal and regarded it as a symbol of luck, the superstitious at that time believed that its power brought you to strange and mysterious places. This idea may have originated because of the mineral's association with the god Mercury, who served to guide lost souls through the kingdom of the dead.

Q. *Where is aluminum found? You don't see it in nature like gold or copper.*

A. You are right in that aluminum is not very visible in nature even though it is the third most abundant element in Earth's crust, after oxygen and silicon. Such common and ubiquitous minerals as feldspar contain plenty of aluminum, but we do not have the technology to extract it as a pure metal. Nature, however, does a lot of the preparatory work for us in the processing of aluminum. Through deep chemical weathering, nature removes such elements as silicon and iron and leaves behind a highly aluminous weathered clay called *bauxite*. It is this bauxite, found mostly in tropical environments, that constitutes the aluminum ore.

This clay is mixed in with cryolite (a fluoride mineral) and heated. It is an electrolytic process: the box or container acts as the cathode and attracts pure aluminum to it while oxygen, the other component of bauxite, is drawn to rods (the anode) placed in the container where it is consumed. This technique is called the Hall Process after Charles Hall. Hall was a student at Oberlin College (Ohio) in the 1880s and developed the process there at age twenty-three. He became wealthy and gave Oberlin

$5 million at his death. There is a statue of him on campus, made of aluminum, of course.

As you know, aluminum is a light, strong metal used in construction and a myriad of other things from aircraft to beer barrels, and in the kitchen as aluminum wrap. The United States and Canada are the leading producers of aluminum. We should mention that at the same time Hall developed his process, it was also being discovered by the Frenchman, Paul Héroult.

Q. *Big land dinosaurs have been found on more than one continent. How could they get across the oceans since I presume they could not swim?*

A. It seems like a baffling question at first, but there are several explanations. We would assume that a particular land dinosaur would not arise on different continents, as that would not be biologically viable. So we are dealing with dispersion from one place. One explanation is that in the past, and still going on in the present, the continents shifted or drifted, and at times have abutted each other. Equally plausible is that a lowering of general sea level created land bridges for dinosaurs and other animals to cross. This is no doubt the way that human nomadic groups came into North America from Asia, across the Bering Strait bridge during the last Ice Age (the Pleistocene).

There are also the factors of time and probability in producing the "rare event." Given millions of years to work with, sooner or later, it might happen that a storm carries a small dinosaur group on a natural raft of trees and vegetation to another continent. This is in the same category as the idea that a troop of monkeys banging at typewriters might produce all of Shakespeare's plays if given many millions of years to keep typing at random. Easier to explain is the spread of plant life whose seeds can be dispersed long distances by winds or by birds. That is how vegetation came to newly risen volcanic islands in past centuries in the Atlantic and Pacific Oceans.

Q. Is it now accepted as proven fact that an asteroid crashing into Earth caused the extinction of the dinosaurs?

A. No. It is an attractive theory. The dinosaurs became extinct 60 million years ago at the end of the Cretaceous period. This extinction included reptilian groups in the world's oceans as well as land dinosaurs such as Tyrannosaurus. Scientists have found that Late Cretaceous rocks contain an unusually high content of iridium, a metallic element resembling platinum. It is somewhat rare on Earth but is fairly abundant in asteroids. We also know that asteroids do indeed crash upon Earth, as Meteor Crater in Arizona testifies. Given an asteroid large enough, the collision would be a catastrophe, darkening the skies with iridium-laden dust, shutting out sunlight, and turning the entire planet colder. Growth of vegetation, a necessary food-chain item, would be arrested. Yes, it could have happened. Those who favor this theory believe that asteroidal bombardment can also explain periodic extinctions at other times in Earth's geologic history. They postulate that the Sun is really a double star, having a dark stellar companion that returns every 26 million years or so to launch another asteroid at Earth, causing extinctions at that time. It seems to be a viable theory.

Still, there are unanswered questions. Some paleontologists maintain that the dinosaurs had already started to become extinct 10 million years before the asteroid arrived. There is a strong correlation between extinctions, large and small, and the rise and fall of worldwide sea level as reflected in the geologic record. So far, no double-star companion of the Sun has been found. Is asteroid bombardment the only way nature causes a species to die out? Better answers await more evidence.

Q. *How did scientists think the dinosaurs became extinct before the asteroid theory came along?*

A. There were many ideas offered, most of them not very good. Mammals ate the eggs of dinosaurs so they failed to perpetuate; the dinosaurs caught some kind of disease that wiped them out; or, dinosaurs had a small brain and were too dumb to survive (after surviving for 200 million years). None of these ideas have much merit. Any sound theory must take into account reptile extinction in the sea as well as on land. It should even explain why some reptiles survived while others died out. There should be a mechanism with known or plausibly expectable effects. For most geologists, that mechanism was the rise and fall of sea level.

These sea changes are imprinted on sedimentary rocks extending back more than 600 million years and can be deciphered using standard, well-tested geologic field methods. Even the duration of these events can be determined. Whether sea level rises or falls, it ushers in profound changes in climate, land distribution, and ocean depths. This in turn affects vegetation, temperatures, rainfall, and food supply. Subenvironments are destroyed or modified to a more or lesser degree, and new ecological niches appear. Dire or benevolent effects on the occupants of each environment is inevitable. That mass extinctions occur harmonizes well with major withdrawal of seas at the end of geologic eras, including the Cretaceous (end of the Mesozoic era). It can be noted also that certain life-forms at such times spread and became more dominant. Mammals and angiosperms (flowering plants) are such examples, following the Cretaceous when the asteroid presumably struck. This concept still has considerable merit in explaining extinctions of the past.

Q. *How high would sea level rise if all the ice at the polar caps melted?*

A. It is estimated that there would be a worldwide rise in sea level of perhaps 200 feet. If it happened with great rapidity, such as over a fifty-year period, it would be an unmitigated disaster. The heaviest populations live on or near the world's shorelines and these would come under water, forcing migrations inland. The Florida Peninsula would be erased, and marine life would move in where humans once lived. Weather patterns would change drastically. Many rivers large and small would turn into estuaries and lakes. It would be a new and probably dangerous world as humanity packed themselves into a much-reduced living space. Original inhabitants would resent the refugees of the inundated land and the likelihood of violence, disease, and war would increase. Nations would make war simply to acquire livable land. Much farmland would be destroyed, resulting in acute food shortages.

It is doubtful if this all could happen. However, in actuality the sea level is rising, but at a rate of inches per century, and this is because the polar areas are feeling the result of incipient global warming. If the sea encroachment is slow enough, the world may be able to adjust and prepare remedial measures. We hope.

Q. *Is it true that people in Siberia dine regularly on frozen mammoth meat?*

A. The inhabitants of Siberia do not dine regularly on mammoth meat. Some have tried it and found it would hardly receive the label of "choice." Although it is really not fit for human consumption, dogs and other scavengers will tear at the meat and bone when thawing and erosion exposes these extinct beasts. Primitive man hunted mammoths as late as 8,000 years ago, which is fairly recent, geologically speaking. Tribes would cooperate in driving a mammoth off a cliff and share the meat, skin, and bone, for all parts were utilized. It is improbable that humans caused their extinction. The exact reason is not known, but it is suspected that climatic change following the last Ice Age changed the vegetation enough to reduce the mammoths to extinction.

Q. *I saw a movie about a caveman frozen in ice and thought of the mammoths in Alaska and Siberia. Why aren't more humans found frozen in ice?*

A. A popular misconception is that the mammoths are frozen in ice. They are not. They are found in frozen sediment of clay, sand, and silt. Such preservation is called *refrigeration* or *entire preservation* by paleontologists. Some mammoths appear to have fallen into ice crevasses, died, and were covered by sediment as the ice retreated and meltwaters buried them under its sediment load. Others fell through ice while crossing a river and were swept by the current to calmer waters where they were entombed by river sediment. Others may have been the victims of mud avalanches.

The preservation of humans by like means is much rarer. Humans avoided the common traps that victimized mammoths because they were lighter, more agile, and smarter. They probably had better eyesight, too. But it has happened. On St. Lawrence Island in the Bering Sea the body of an Eskimo woman was found who was the apparent victim of a landslide 1,600 years ago. The body was in such a perfect state of preservation that scientists determined that the woman suffered from arteriosclerosis. Then there is the famous "Iceman" of Europe. Examining scientists presume that he froze to death in a blizzard while crossing high mountains near the present Italian border. His body, in excellent preservation, was found with all of his clothing and possessions intact.

Q. *What is the difference between a mammoth and a mastodon?*

A. There is very little. Both are members of the elephant family and there are minor differences in the skull, tusks, and teeth. The tusks are actually modifications of incisor teeth. From the fossil record, it looks like the mastodon arrived on the scene first, and then developed a side branch that led to the mammoth. The mammoth in turn gave rise to the modern elephant. Both the mastodon and mammoth survived up to 8,000 years ago and coexisted with man. There were tremendous herds of these animals. Their ivory tusks have been exported from Siberia since the Middle Ages. The environmental changes that affected their food supply, or other survival factors, must have occurred relatively rapidly. Nonetheless, their bones and tusks will still be around for a long time.

4

A MOBILE EARTH

Q. *Do scientists still believe that continents drift around?*

A. Yes, and it has become rather established fact. The concept goes back as far as the sixteenth century when it could be seen from world maps that Africa and South America could be fitted together like two pieces of a puzzle. But it was not until 1912 that Alfred Wegener, a German meteorologist, drew greater attention to the idea. He believed that all the continents were at one time joined together into one supercontinent. He visualized it breaking up into pieces perhaps 200 million years ago, before slowly drifting to their present positions. Most scientists rejected the drift theory because they could see no force or mechanism to drive the continents from place to place. The theory remained dormant for more than thirty years until a series of discoveries showed that Wagener was right after all. Meanwhile, Wagener himself undertook an expedition into the Greenland Ice Cap at age fifty, and was never heard from again.

Q. *I see more and more mention of tectonics and plate tectonics in magazines and newspapers. Don't they have something to do with earthquakes and volcanoes?*

A. They have a lot to do with both of them. The term *tectonics* refers to the deformation of Earth's crust, best exhibited in the layered sedimentary rocks that cover most of the continents. Here we see seashell fossils in layered rocks atop mountains, clear evidence that the ancient sea floor has been uplifted several hundred or even thousands of feet. There are layers that are folded and others that are broken and displaced by faulting. Tectonic forces build mountain ranges such as the Rockies and the Andes. There are also broad and gentle downwarpings and upwarpings of the crust. All of this points to the fact that Earth has been, instead of being an inert ball of rock, an active and dynamic planet undergoing continual change in response to powerful tectonic forces within. That this process is still going on is attested to by modern-day earthquakes and volcanic eruptions.

As to *plate tectonics*, this is the current theory as to how and why deformation, earthquakes, and volcanoes occur. It was from the old theory of continental drift propounded by Wagener that the concept of plate tectonics evolved. In brief, Earth's crust is seen as made of twelve or more plates, or huge rigid slabs, that might resemble an egg whose shell has been cracked in several places. However, these plates can move. They may be found moving away from each other (divergent plates), bumping into one another (converging plates), or sliding past each other (transform plates). It is along the margins of these plates that we find increased volcanic and earthquake activity. There is excellent evidence that along midoceanic ridges, there is upwelling of upper-mantle material in response to heat, which spreads out laterally, creating new seafloor. In the process, the plates move and carry the continents, perched on the plates, with them.

Q. *If the seafloor is spreading, adding new crust to the earth, why isn't it expanding like blowing up a balloon?*

A. Scientists pondered this exact question. In studies of tectonic plate behavior, it was found that when two plates meet, one will sometimes duck under the other one and be reabsorbed into the mantle. A good example is Japan, where the Pacific plate is ducking under Japan. This process is known as *subduction,* and shows that new crust is balanced by elimination of old crust. Along subduction zones, deep oceanic trenches also develop. An interesting aspect of plate tectonics is that the continents themselves do not become subducted because they are very light and cannot be pulled down into the subduction zone. We see confirmation of this in that the seafloor is very young, geologically speaking (not further back than the Mesozoic era), while continental rocks are as much as 3 billion years old.

The most widely held idea is that the driving mechanism for plate tectonics is the convection currents in the mantle below the crust that bring new mantle material to the surface and cause seafloor spreading. The convection currents also assist in dragging older crust back into the mantle at subduction zones. An analogy is often made of a pan of tomato soup simmering on a stove top where the soup rises to the surface, cools a little, and descends only to come back up when it heats up again. Without heat, of course, there would be no plate-tectonics activity. It is thought that the source of the heat is from residual heat of Earth's beginning and that generated by radioactive decay.

Q. *I heard somebody mention "the ring of fire" around the Pacific Ocean. What does this refer to?*

A. If on a map you trace the general border area of the Pacific, you will see that it includes Japan and the Indonesia area, both noted for volcanoes and earthquakes. It also includes volcanic islands across the South Pacific, and then north to the coast of California and Alaska, also known for their earthquakes and volcanoes. This concentration of earthquake and volcanic activity in the Pacific rim has earned it the title of *ring of fire*.

What is happening is multiple tectonic-plate contact. The Pacific plate is moving westward and being subducted under Japan. There is also the relative movement of the Pacific plate moving northward while the North American plate moves south, which has created the San Andreas and other faults along the West Coast. At the same time, pressure is exerted against Alaska, known for its severe earthquakes. It should also be pointed out that local *hot spots* could be found within and under a plate. Such a hot spot is responsible for the volcanic activity that formed the Hawaiian Islands, which do not lie on a plate margin.

Q. *Just how fast do continents move as they drift?*

A. Very slowly, often only about 2.5 centimeters per year, which is about an inch. So it would take a century to move 8 feet. That seems almost insignificant in the time and distance frames we are used to working with. But if we alter the time scale to realize that here nature is using millions of years at a time to accomplish something, the picture changes. In one million years, continents, or plates, could move 20 miles. Since the "recent" Ice Age began two million years ago, the plates have moved 40 miles. Since the dinosaurs became extinct, continents would have shifted around more than 1,000 miles. A map of the geography at that time would be almost unrecognizable. There is value in this information of continental movements across the face of the globe. It explains, for example, why we have ancient tropical swamps making coal beds in Pennsylvania millions of years ago. The region that is now Pennsylvania was drifting through the equatorial zone.

Q. *How can rocks become tilted and even overturned?*

A. Yes, rocks of Earth's crust have been overturned in many places, especially where tectonic action takes place, but it normally takes hundreds of thousands or millions of years, and cannot happen while you watch. To see what may happen, imagine a sequence of layered sedimentary rock being subjected to lateral compression from either side, a squeezing effect. The application of force is so slow the rocks become plastic and flow like sealing wax rather than shatter like peanut brittle. An upfold, called an *anticline*, is formed such that it builds higher and higher. It may tend to tip over and lie down in a recumbent position. Some of the rocks that make up this fold have been rotated more than 90 degrees and are upside down. You can illustrate this yourself with a sheet of paper or the blanket on your bed.

We also see large sheets of rocks that have likewise been squeezed, but broke along a low-angle plane where movement can take place. Great displacement along this plane is called *thrust faulting*. The overthrust sheet can override itself, and be thrusted many miles. Geologists think of this as crustal shortening. It can take millions of years, but examples are known in the Rockies and Appalachians.

Q. *Is it true that Earth's crust is elastic like a rubber band?*

A. Perhaps Earth's crust is not that snappy, but it does move up and down. Across all the continents we see layers of solidified sediment that could only have formed on the ocean floor, and yet are now on the land, many hundreds or thousands of feet above sea level. A striking example is what has taken place during the past two million years since the Ice Age. Great ice sheets of ponderous weight engulfed the continents and shoved them down. When the ice retreated and melted, the continents rebounded elastically. The reason we know this is true is that old lake shorelines are at different levels where they should be at the same level, indicating that some parts of Earth's crust rebounded faster than others. Some of these old shorelines are even tilted, and water cannot sit on a slope. Today, the sediment being deposited at the delta of the Mississippi River is considerable, as the river drains half of a continent. Careful measurements show that Earth's crust in southern Louisiana is sinking down, but very slowly.

Q. *Couldn't Moses' crossing of the Red Sea be explained by continental drift? The Red Sea would have been much narrower then.*

A. The "drift" just isn't that fast. There is little question today among geologists that continental drift does indeed occur, and furthermore, that the Red Sea is a fine example of rifting apart from the African continent. If we assume that the Exodus took place about 3,500 years ago (more if you like), the separation of the Arabian lands from Africa to form the Red Sea would not have increased during this time span by more than a few hundred feet, maybe less. The Red Sea is 100 miles wide in places. The idea that walls of water were rolled back to create a highway such as depicted in Hollywood movies cannot be thought of as accurate history. More likely, Moses crossed the Gulf of Suez at a place of very low water. Some have speculated that the Exodus coincided with the violent eruption of the volcano at Thera, which caused giant sea waves to slosh around the Mediterranean, temporarily withdrawing waters from the Gulf of Suez, which later returned in time to drown the army of the pharaoh. We can't know for sure.

Q. *How long did it take to carve out the Grand Canyon?*

A. The Grand Canyon was cut by running water of the Colorado River, and it took, by the estimates of geologists, about 15 million years. It is hard to believe that just water and its sediment load, running over rocks, could do this, but given enough time, it can and does. We are reminded of the saying of a geologist about the mythical little bird that comes once a year to sharpen his beak by rubbing it against a great rock; in several million years, the rock is no more. About 15 million years ago, the Colorado plateau was subject to a swift elevation of Earth's crust. This quickened the old meandering Colorado River in its course, which with renewed energy cut down to create this wonder of the world. As the Colorado River continues its work of downcutting, it is flowing through some of the oldest rocks on Earth.

Q. *Do scientists agree that San Francisco could fall into the sea if another big earthquake occurs there?*

A. The famous San Francisco earthquake of 1906 was caused by movement, in some places as much as 18 feet, along the San Andreas Fault. However, most of this movement is sideways or laterally without much up-and-down movement, as with other kinds of faults. Also, the site of the city is on part of the continental plate, and therefore is of lighter rock material than the denser basaltic material of the Pacific plate. In short, it is hard to sink. Geologists recognize that there is now enough pent-up force, a lock on the fault, in the San Andreas to cause another severe earthquake should movement occur. We doubt seriously if the city would be thrust beneath the sea during an earthquake, although there might be some minor flooding in places. The concern is that San Francisco is a much bigger, more densely populated city now than it was in 1906.

Q. *What is the worst earthquake on record?*

A. It depends on what is meant by "worst." In terms of loss of life, the most disastrous quake we know of was in Shen-Shu, China, in 1556. An estimated 830,000 people died. More recently, and again in China, a quake of 7.7 on the Richter scale took place in 1976, devastating Hebei Province and killing an estimated 800,000 people. Compare the scope of these calamities with the highly publicized San Francisco earthquake of 1906, which killed around 700 people.

On the other hand, we can speak of the quake in terms of how much energy was released. Since measuring instruments first came into use, there have been several major earthquakes exceeding 8 on the Richter scale. One quake we think might have been the most intense was that in Lisbon, Portugal, in 1755. The ground shook for six minutes at one point and the city was completely destroyed. It took nearly a week to put out all the fires that started. Crewmen aboard ships in the harbor were tossed into the air. Large sea waves, called *tsunami*, reached heights of 40 feet, sweeping over the city. These waves traveled across the Atlantic and were still 12 feet high when they reached the island of Martinique in the Caribbean Sea. Finally, one can speak of an earthquake in economic terms: the loss of property, goods, and services.

The earthquakes in the late twentieth century in California alone cost billions of dollars.

Q. *An earthquake in Idaho made a lake appear. How was that possible?*

A. In a sense an earthquake reshapes part of our landscape. When movement along a fault occurs, you get an earthquake. This can result in the shifting of the land surface as much as several feet at a time—up, down, or sideways. For example, after the San Francisco quake of 1906, it could be seen that fence lines that crossed the fault zone had been offset six to ten feet or more. Suppose that instead of a fence line, a stream had been offset, moving higher ground across its course and blocking the normal flow of water. The stream water would back up and impound like water behind a dam. This is how Bear Lake in Idaho formed. Another possibility is that the wrenching movement of the earth during an earthquake could trigger a landslide into a stream valley, and block the flow of water that way. This is what happened in West Yellowstone back in the 1950s. The lake that formed at that time is still there and is appropriately called Earthquake Lake.

Q. *How do scientists know that a particular region was subject to earthquakes in the distant past?*

A. If it is not too distant in the past, then they know. Damage from earthquakes may appear in buildings and other structures. For example, the quake that struck Charleston, South Carolina, in the last century caused building damage that can still be seen today. However, even in areas without buildings, geologists would look for evidence of faulting in the rocks of the area. A *fault* is a crack or fissure in the earth along which rock movement has taken place accompanied by the enormous release of energy that causes earthquakes. The displacement of traceable rock units can be measured. If there is considerable displacement of tens or hundreds of feet, that would suggest several strong earthquakes having occurred, even millions of years ago. There are hundreds of thousands of faults around the world. Fortunately, not all of them are active now.

Q. *Will scientists be able to predict when and where an earthquake will take place? It would save a lot of lives.*

A. We can certainly point to earthquake-prone areas such as along the San Andreas fault in California to answer the "where" of your question, although there are occasional surprises as when a severe quake struck New Madrid, Missouri, in 1912. This area was not considered earthquake-prone at all. So, even predicting where a major tremor will occur is an art still developing. It is the "when" of an earthquake that is of critical importance. So far, it has been learned that there are certain clues that foreshadow a major quake. The pressure buildup along a fault will start to open new fractures into which groundwater migrates. The groundwater contains radon, a short-lived radioisotope. Its appearance in water wells may be a signal. At the same time, sensitive tiltmeters deployed at strategic locations can detect slight upward movements of the ground. It has also been suggested that seismic waves from distant sources passing through the suspected area have a tendency to slow down.

Should earthquake prediction reach a practical level, another problem will arise, and that is human behavior. If we were to say, for example, that a major quake would hit Los Angeles in three months, what would happen to property values? Will there be a mass exodus of people quitting jobs and taking all their possessions with them? The impact of several million refugees flooding into the perimeter outside of the vulnerable area would be felt even by those not worried about the earthquake striking their backyard. To make the situation more problematic, supposing instead of a three-month lead time, it was less than a week. This would cause panic, as when the cry of "Fire!" is heard in a crowded theater. These are social questions we may have to confront some day in the future.

Q. *How deep in the earth is the lava that comes out of volcanoes?*

A. Pretty deep. Hawaiian volcanoes are among the most closely studied and monitored in the world, and we know quite a bit about them. During the eruption of Kilauea in 1959–1960, the movement of the magma upward from below could be traced by a sequence of small earthquakes that accompanied the movement. Geologists think the molten material originated 30 miles deep in a weak, plastic rock zone known as the *asthenosphere*. This zone lies within the upper mantle and would be considered a "hot spot" because Hawaii lies within the Pacific plate. The molten lava eventually reaches the surface, with spectacular flows and lava fountains. The Hawaiian Islands are among the largest and tallest volcanic edifices in the world, but most of it is, of course, under the sea. This lava is mainly dark basaltic lava. Some of it may originate even deeper.

Q. *I read that there are many hot springs in Iceland. How can a place so cold have hot springs?*

A. Hot springs can be found around the world in places such as New Zealand and Yellowstone Park in the western United States. Their climates are different from Iceland. What is interesting is that the hot springs and what lies beneath them gave birth to a land as cold as Iceland. These waters come from molten-rock sources far below the surface that provide the raw material ejected from volcanoes. In the remote past, eruption of molten-rock material on the ocean floor caused the buildup of basalt, eventually reaching the surface and forming the island we call Iceland. The molten magma and gases are still stewing beneath Iceland. The emergence of the volcano Surtsey during the 1960s south of Iceland shows how land is formed. Our fiftieth state, Hawaii, was formed in a similar fashion as well as most of the islands in the Pacific.

Q. *Is there any chance Mount St. Helens will erupt again soon?*

A. There is always that possibility, although usually an eruptive phase is followed by a long period of dormancy. When Mount St. Helens erupted on May 18, 1980, it was the first time it had done so in 123 years. Mt. St. Helens is only one of a string of volcanic peaks along the Cascade Range that has the potential to erupt. Among these are Mt. Shasta, Mt. Rainier, Lassen Peak, Mt. Hood, and Mt. Baker. Lassen Peak last erupted in 1915. Geologists have found that these are actually newer volcanoes built upon the remains of much older volcanoes dating back more than 50 million years. The "new" volcanoes like Mount St. Helens are only one million years old.

Q. *What was the world's biggest volcanic explosion known?*

A. A leading candidate is Krakatoa in the East Indies, erupting with unparalleled violence over a two-day period in August 1883. All through that summer, the three small cones that comprised Krakatoa were smoking and steaming ominously. Then on August 26, the eruption began with a loud explosion heard 100 miles away and a cubic mile of pulverized volcanic debris (ash, cinders, and pumice) was thrown 17 miles into the air. The next day, the eruption reached its apex with explosions loud enough to be heard in Australia, 3,000 miles away, and volcanic dust turning the skies over Java and Sumatra as dark as night. The shaking of the seabed generated a tsunami, or large sea wave, that reached 120 feet in height, which crashed into the island of Java, killing 36,000 persons. The volcanic dust and ash from this explosion was carried around Earth and took years to settle.

More remote in history and thus having fewer accounts was the eruption of Thera in the Mediterranean Sea area about 3,500 years ago. It was a larger explosion than Krakatoa, if we are to judge by the size of the crater that remains to this day. The blast darkened the entire Mediterranean, causing floods, earthquakes, and rains of glowing volcanic ash. It was a major cause, according to some archaeologists, of the decline of the Minoan civilization on Crete.

5

ICE, LANDSLIDES, FLOODS, AND OTHER ANNOYANCES

Q. *What is the longest river in the world?*

A. The Nile is the longest river in the world. Its waters gather near the equator and Lake Victoria in Africa and flow north through Uganda, Sudan, and Egypt where it empties into the Mediterranean Sea. If we could place the Nile River in the United States, it would extend from New York City to Los Angeles, California, and even beyond. As the river passes through Egypt, it forms a flat-bottomed valley five to ten miles wide, flanked by waterless deserts on either side. It was along this watercourse that the early Egyptians settled between 10,000 to 5,000 years ago, and built one of the great civilizations on this planet. Rivers have provided water for man and beast, and food to eat. They have been the easiest, sometimes the only, way for explorers, settlers, traders, and conquerors to reach other places and peoples. That is why so many early civilizations settled around rivers and ports.

Running water as in rivers and streams, is the greatest geologic agent operative at the earth's surface. A geologic agent is any natural process or force that serves to erode, transport, and deposit rock and sedimentary material. Other geologic agents besides running water include groundwater, glaciers, wind, and waves. The landscape you see around you, with its hills and valleys, whether encountered near home or in your travels, has mostly been sculptured by the erosional and depositional activity of rivers and streams.

Q. *Why is the landscape so varied in one place and flat in another?*

A. In most areas, the topography or configuration of the land is dominated by the erosion and deposition of numerous streams, large and small, and their tributaries. Even in deserts where there is little rainfall and streams flow only part of the year, the landscape is mainly fashioned by streams. They become so torrential when it does finally rain, that they erode in hours what streams in temperate areas would require years to do. The ultimate goal of these geologic agents is *gradation*—the wearing down and leveling of the land to sea level, at which point their energy is expended. Streams achieving this goal form a flat region, called a *peneplain*.

One might wonder, then, if streams have been operative for millions of years, why all the land is not flat. This would be so if not for tectonic activity, or earth movements, that continue to push up the land to create more high areas to be eroded anew. Uplifted plateaus are new high areas where erosive activity has not yet taken effect. It is a cycle that has been operative for millennia.

Q. *Are there other whirlpools around the world besides the one at Niagara Falls?*

A. There are several notable places along coastlines and in the world's oceans where whirlpools may be found. A whirlpool can be generated in coastal areas such as those of Scotland, Norway, and Japan, owing to the shape of the shoreline, its bottom topography, and the influence of incoming and outgoing tides. In river bends, an upward swirling current of water, called a kolk, may occur. In the open ocean, they are simply large-scale eddies. The Sargasso Sea is one. But the notion of a seaweed-clogged area holding trapped derelict ships from Spanish galleons at its center and more recent vintage vessels in the perimeter, is pure fiction. Indeed, if you were on a ship passing through the Sargasso Sea, the fact that it is a whirlpool might have to be pointed out to you. We know of no case where the water swirls down, sucking objects along with it, like that in your kitchen sink. The Niagara Falls Whirlpool is a majestically rotating eddy. You have to watch it for a while to discern the motion.

Q. *What is the highest waterfall in the world?*

A. This would be Angel Falls in Venezuela where the water drops 2,650 feet. Apart from the height, a waterfall may have other equally impressive features. For example, the volume of water passing over Khone Falls, a relatively small falls in Laos, has been estimated at 410,000 cubic feet per second. Or, for grandeur, Victoria Falls in Zambezi with its one-mile width and 17 to 24 million cubic feet of water discharge per second, has been described as the largest curtain of falling water in the world.

Waterfalls and the less steep drops that form cascades and rapids are relatively temporary features of a river course, which eventually smoothes out through its erosive power. Rivers with waterfalls and rapids are considered relatively young streams. Waterfalls form for a number of reasons. Differences in hardness of underlying rock may result in steepening where the stream cuts faster through softer rock. Faulting may displace softer rock across a stream's path and help to create a waterfall. Another type of waterfall is known as the *hanging valley*. This is where a main stream course is greatly deepened by glacial ice scouring the valley. Smaller tributaries are left behind at a higher elevation, and their waters are perforce obliged to drop some distance to the valley floor. The Upper Falls in Yosemite Park, with a drop of over 1,400 feet, originated this way.

Q. *Is there any scientific support for the Flood that prompted Noah's Ark to house the only life left on Earth?*

A. More than one billion cubic miles of water would have to be added to the oceans to cover all of Earth's mountains. Science knows of no source for such a vast amount of water, or where it would have gone when the flood receded. There is also the problem of space in the ark. You would need to house more than forty-three thousand animals. The ark wasn't that spacious. Eight people—unskilled zookeepers—would have to care for these animals for nearly a year and provide food and water. Despite thorough searches of Mount Ararat, the presumed docking place of the ark, no convincing evidence of its remains have been found. It should be noted also that a pair of animals, male and female, is insufficient to insure the continuation of a species. Most biologists would agree you would need a herd of thirty or so mixed individuals for the species to survive. These are only a few of a multitude of rational objections to the literal veracity of this account.

If the story were in fact true, then all humans are descended from Noah's family, including such diverse peoples as African Pygmies, blond Swedes, and Alaskan Eskimos. Big but local floods did occur in Noah's geographic area; science has evidence of them. This may have been the basis for the story of the flood as well as earlier large floods written about in Babylonian and other ancient texts.

Q. *Some horror movies show the villain getting sucked down to his death by quicksand. What is so special about this sand and where is it mostly found?*

A. First of all, there is no special kind of sand involved, although at one time it was thought so. Any sand can become "quick." When you are walking along a sandy beach, the sand supports your weight because all of the sand grains are in mutual contact and have a bearing strength. However, if there is an upward moving current of water passing through the sand such as from a spring, then the grains lose contact with each other and the sand-mud-water mixture behaves like a liquid. The body of an animal or human would then start to sink.

Quicksand does not suck you down, contrary to folklore. The sand-water suspension actually has more buoyancy than water alone because it is denser, and according to physical laws, you are buoyed up by a force equal to the weight of the liquid displaced. The reason why animals and some humans have drowned in quicksand is the fear, panic, and ensuing struggle to stay above the quicksand. Quicksand is more often found in swampy regions because they are areas of possible intersection with the groundwater table, and thus have a greater likelihood of rising waters from below. It is also possible that an impermeable layer of clay or similar material may prevent effective drainage, so the water above the clay becomes saturated. Vibrations of the ground such as in an earthquake may cause sand grains in water to spring apart, creating a quick condition, and adding to the hazards of quakes. If caught in quicksand, remain calm and yell for help.

Q. *Can it be true that a landslide roars down a slope with great speed because it does not touch the ground?*

A. Oddly enough, it is not only true, but you can verify it for yourself. If you drop a book on a table, notice that air escapes out from under the book and disturbs light objects nearby. In the same way, a mass of millions of tons of rocks and soil detached from above and moving down a slope will cause air to rush out from under it at velocities of 50 or 60 mph. There will be a cushioning effect by the air before it can escape. Thus, the landslide may be riding on a layer of compressed air a few inches above the surface of the ground. This also accounts for the speed at which landslides can move—up to 100 mph—because of the lessened friction. The proof of this is in careful examination of the track of the landslide in the aftermath. There will be delicate blades of grass and small pebbles that remain undisturbed. This was observed in the case of the Hebgen Lake earthquake and landslide in Montana, August 1959, when twenty-eight people died, most of them campers in Madison River Canyon. It had to be a terrifying experience as the earthquake took place during the night, triggering a huge landslide, and spilling torrents of water over the Hebgen Lake dam. Thus, with great rapidity, there was a surging flood capable of moving automobiles, 50 mph winds, and a tremendous roar of the wind issuing from the darkness. Fault scarps appeared and the ground dropped several feet. It was another example of how savage the earth could be.

Q. *Why does the Leaning Tower of Pisa lean? And why hasn't it fallen over?*

A. Originally designed to be a bell tower, the Leaning Tower construction began in 1173 under the direction of Bonanno Pisano, but was not finished until two hundred years later. The lean became obvious when the height reached 35 feet. The reason for the lean was the poor footing of the foundation. Although the workers had dug down 10 feet, they found no bedrock, only sandy soil differing in bearing strength. As it was built higher, they added stories at an angle to straighten the tower out. Nonetheless, at its present height of 185 feet and seven levels, it leans out 17 feet from the perpendicular. The tower has walls eight feet thick and a diameter of about 50 feet. Yet the lean keeps increasing, measuring in millimeters.

During the 1960s cement was added to the foundation but this proved ineffective. In 1989 the tower was closed to the public and $4 million was earmarked for foundation work to prevent collapse of the tower. This new effort at preservation began in 2000. One might wonder at all the fuss over one little tower. However, the Tower of Pisa is an outstanding example of the architecture of the Middle Ages and is, after all, eight hundred years old. It is also a popular tourist attraction. In the year of its closing, it drew 700,000 visitors who climbed to the top floor. It should be hoped that somehow engineers can keep the tower leaning without danger of collapse. It is part of its charm.

Q. *Were the Roman catacombs that the Christians hid in natural formations or were they dug out?*

A. The catacombs were dug out by human effort from layers of suitably soft sedimentary rock material, such as volcanic tuff or other volcanic rock. Their purpose was to be used as underground cemeteries and were not expressly dug as hideaways for Christians during the Roman persecutions. In fact, Jews also constructed catacombs. Similar sites can be found in Egypt, Malta, Sicily, and Tunisia but the ones in Rome are the most extensive, elaborate, and best known. The catacombs are a mazelike series of passageways, galleries, and burial niches descending to various levels. The burials are simple, cloth-wrapped bundles without coffins. Besides burials, the catacombs were occasionally used for funereal feasts and prayer meetings.

The catacombs were first dug as family burial vaults, then grew as friends and fellow Christians were also received underground. Since they already existed during the time of the persecutions, they were logical places to seek refuge. They served the same purpose during the barbarian invasions such as those of the Goths in later centuries. With the end of Roman persecutions by the edict of Milan in 313, the catacombs were largely abandoned, the entrances covered by erosion and often forgotten. They were rediscovered in the seventeenth century, studied, and mapped. Many passageways remain to be explored. Some large chambers such as those near the Appian Way are now open to visitors.

Q. *How can a solid rock just turn into clay?*

A. It is a perfectly natural process called *chemical weathering*. It goes on all the time. However, it is a remarkably slow process taking thousands, even millions of years. You might think of the earth's surface as a gigantic chemical laboratory where the minerals of rocks, which are chemical compounds, are reacting with water, oxygen, and carbon dioxide in the atmosphere. The result is rock decay. The process is little different from the rusting of iron objects left out-of-doors, for example. The process is hastened in the presence of warmth and moisture. Such weathering can also take place deep underwater, sometimes with the aid of small organisms. Scientists, who have studied metal from the wreck of the ocean liner *Titanic*, believe that in another century or two, the great ship will be only a pile of rust. Incidentally, the moon rocks brought back by astronauts contain very fresh-looking minerals, even though they are of great antiquity. This is because the Moon has no atmosphere, and so no natural chemical laboratory.

Q. *What is the largest lake in the world?*

A. There are both fresh-water and salt-water lakes. Among salt-water lakes, the Caspian Sea, with its 4,000 miles of shoreline between Europe and Asia, qualifies as the biggest lake in the world. If we are talking about fresh-water lakes, Lake Superior is the largest lake in terms of surface area, but it does not contain nearly the amount of water as in Lake Baikal, a very deep lake in Siberia containing 5,500 cubic miles of water.

Lakes represent a temporary inland basin, or base level, for waters journeying to the sea within the constraints of the hydrologic cycle. Within that cycle, lakes are quite minor, comprising only 0.4 percent of total water, but are of critical importance in human use. We use lakes for drinking water, power generation, transport, commerce, and recreation, amid a myriad of functions. Most of the present lakes of the world are the result of meltwaters remaining when the ice retreated at the conclusion of the Ice Age 12,000 years ago. In North America, the Great Lakes and the millions of smaller lakes across the United States and Canada are the vestiges of the great Wisconsin ice sheet that covered the area. Alaska alone contains three million lakes.

Q. *Why is the Dead Sea so named?*

A. Although it has other names, for example, the Salt Sea and the Sea of Zoar, the 50-mile-long Dead Sea is probably the most descriptive name for the lake into which the Jordan

River flows. Its waters have a salt content of 25 percent, so that fish are not able to live there. Some bacteria are the only resident occupants of the lake. Being the lowest body of water on the earth's surface at 1,300 feet below sea level, the Dead Sea has no outflow and instead there is evaporation as new waters are added from the Jordan and other rivers as well as from underground springs. When fish and other organisms are swept into the Dead Sea from the Jordan River, they die immediately. The Dead Sea, Sea of Galilee, and the Jordan River lie along a part of the rift system where detachment of East Africa from the rest of the African continent is taking place. Artifacts found by archaeologists indicate there has been human occupation of the area for more than 500,000 years.

Dead Sea evaporite minerals have supported an industrial production of potash and salt, as the basin is an enormous reservoir of saline minerals. Tourists come for the mild winter climate, the historical associations such as Masada, and the caves where the Dead Sea Scrolls were discovered. However, there are no large modern cities anywhere around the shoreline. Moreover, the summers are intensely hot. Many biblical events center upon the Dead Sea, with perhaps the destruction of Sodom and Gomorrah being the best known. The ruins of these cities probably lie in shallow water at the southern end of the Dead Sea. Eventually, the Dead Sea seems doomed. Its waters continue to shrink, brought on in part by bromine-production facilities operated by both Israel and Jordan.

Q. *I recently visited Carlsbad Caverns. Isn't there more to their formation than just dissolving out by plain old water?*

A. Most caves are formed by the dissolution of limestone, but as the water seeps into the ground, it becomes charged with organic acids from plants which makes it a much more effective agent in dissolving limestone. It is not just plain old water, which actually would not work very well at all. The chemical reaction, albeit slow, is between calcium carbonate, the chief ingredient of limestone, and a weak carbonic acid. A small cavity develops. Over a period of thousands or even millions of years, a cave is formed. Eventually, extensive caverns develop such as those found at Carlsbad. If there were no limestone, there would be few caves.

Water dripping from the roof of a cave forms the well-known *stalactites* or "icicles" as the water evaporates and redeposits calcium carbonate. Excess water reaching the floor of the cave builds up the piles of calcium carbonate known as *stalagmites*. There are also effects in the topography on the ground surface above the subterranean limestone. Sinkholes, circular depressions often filled with water, dimple the landscape resulting from collapse of the cavern roof. Some surface streams disappear underground to continue the work of dissolution on a greater scale.

Caves were a boon to early human hunters and gatherers such as the Neanderthals and later the Cro-Magnon, who made them their homes, a refuge from the elements as well as from predators. And it was here on cave walls that we see the famous cave paintings, the early artistic endeavors of early man in France and Spain. Caves are also a boon to archaeologists who explore these caves in search of how these early humans lived and worked, and even what they thought.

Q. *Can water in the ground be located by water witching? Is there a scientific explanation for its success?*

A. This is called *dowsing for water*. A person holds a forked stick, walking over an area where he hopes to find water. When the stick seems to be pulled downward, water in the subsurface is purportedly indicated. In many cases, drilling reveals that water is present. This is not surprising because water should be found anywhere you drill if you go deep enough, even in the Sahara Desert. Groundwater is much more voluminous than surface water. That a stick will locate water is nothing more than a superstition with no scientific basis, though it is believed by many.

A few dowsers who enjoy their reputation will also cheat their gullible followers. They have been known to carry state groundwater maps in their vehicles, and when they find water, they claim it was the stick. Some also claim they can find gold this way, but they do not appear to be very rich. We would suggest a more logical course if you wish to have a well. Consult state or federal agencies that have groundwater maps of your area; they will be happy to advise you as to your prospects, including how deep you may have to drill.

Q. *I wonder how a whole forest could turn to rock like in Arizona.*

A. One of the myths of the Petrified Forest is that the trees found there today actually grew there, and even advertisements for the Petrified Forest depict trees in a standing, growing position. This is false. Geological research shows that the trees actually grew in Colorado about 180 million years ago. They died, fell, and were transported by rivers into the northern Arizona area to be buried in sand and silt. With time, circulating groundwater loaded with silica replaced the woody fibers with a molecule of silica on a volume-for-volume basis, thus preserving the woody character of the trees, even the bark. This is well explained at the park for those willing to listen.

Smaller fragments of petrified wood can be found all over Texas, so it is not a rare phenomenon, although the preservation of entire tree trunks as at the Petrified Forest is unusual. In the geologic calendar, the Petrified

Forest was formed during the Late Triassic and the trees were entombed in the Chinle Formation. This was a time when mammals on Earth were small and insignificant, and the reptiles were starting their ascendancy to domination by the larger dinosaurs.

Q. *Is it true that the ice once covering part of the United States and Canada was in sheets more than a mile thick?*

A. It is true, and in places the ice was more than a mile thick. We know they began to melt and recede about 12,000 years ago. We know the extent of these great ice sheets because during their advance, they eroded and picked up enormous quantities of sand, silt, clay, and even large boulders. When the ice melted, they left this sediment behind on the ground. By mapping this glacial debris, called *glacial till*, we know how far it reached. We have a good idea of the thickness because the ice caused the earth's crust to sag down due to the weight. Great ice sheets still exist in Greenland and at the South Pole, where they are well known to be more than a mile thick.

During the advance of an ice sheet or glacier, it acts like a big bulldozer, indiscriminately sweeping up rock and sediment in its path and carrying this load many miles. When deposited, the till may assume landforms in the shape of sinuous ridges and irregular piles. These are called *moraines*. It is clear that the source of all the water necessary to make these ice sheets was the oceans. This caused significant lowering of sea level, as much as 300 feet, changing the configuration of shorelines around the world, creating land bridges for animal migrations, including the primitive men and women who entered North America from Asia.

Q. *I saw a large mass of rock with parallel sets of scratches that someone said was caused by ice. Isn't this another case of a weak explanation of something we know nothing about?*

A. To the contrary, these scratches or *striations* are well understood by geologists. It is only baffling if you think of ice as cubes in a freezer or a thin layer to slip on while walking on a sidewalk. We see these scratches on bedrock in areas we know were once glaciated. Imagine ice covering this area on a large regional scale and with thicknesses of several thousand feet. We know this to be true because of the great drop in worldwide sea level, the oceans being the only logical source for all the water to make the ice. These ice masses move slowly and plastically and pick up stones, which are gripped by the ice at the base of the ice sheet. As the ice moves along, the stones held by the ice do the scratching. Geologists studying modern glaciers have observed such striations forming. Those formed long ago probably developed the same way. Note also that these scratches provide a clue to the direction of ice movement. Compass bearings taken at a number of such scratch sites can reveal an overall pattern of ice advance.

Q. *I am confused. Some say a new ice age is coming, then I hear about global warming and the greenhouse effect. What is going on?*

A. There is confusion and uncertainty about this. One geology textbook states that in the future the earth will get "hotter or colder." This seems to cover all the bases. Actually, we could still be in the Ice Age, because ice ages are episodic with more than one advance and retreat of continental ice sheets, punctuated with interglacials when things warm up. One episode, the Wisconsin, ended 12,000 years ago, which seems like a long time but it is a geologic yesterday. Thus, we could be heading into a warmer interglacial period, with another major ice advance maybe 100,000 years from now. We simply do not know.

For the time being, the signs favor more warming. There is discernible melting of the Antarctic ice cap and retreat of the big Columbia glacier. Sea level seems to be rising about an inch a century. Add to this the exacerbating effects of human activities in the burning of fossil fuels and increasing the amount of carbon dioxide and methane in the atmosphere, which helps retain heat. Whichever outcome prevails—hotter or colder—it is going to take a long, long time.

Q. *Why have the United States, Denmark, and other countries spent so much money drillings holes in the Greenland ice cap? It seems that money could be better spent where there are people starving around the world.*

A. By drilling through this very thick ice, cores of the ice are obtained for study. The deeper ice is many thousands of years old and can actually be dated because small air bubbles, trapped in the ice, represent the composition of the earlier atmosphere of the earth. Since this air contains carbon dioxide, it can be dated by the carbon-14 method. It tells us how the atmosphere has changed over many thousands of years (it has not changed much, despite our polluting of the air). Also, layers of dust particles represent major volcanic eruptions in the earth's past and this can be related to climatic change. The more we understand Earth and its processes, the better we are able to control the forces of nature for the benefit of humankind. Therefore, it is not a waste of money. Also, it is good to see nations cooperating on a worthwhile project rather than making war on one another.

Q. *I was studying a map of the world and noticed that big deserts form a band around the earth north and* *south of the equator at about 20 degrees latitude. Is this just a coincidence?*

A. No. This pattern results from winds moving from higher latitudes toward the equator. They pick up moisture as they go, but are unable to release it as rain. This is because as they get closer to the equator, they increase in temperature and thus can absorb greater amounts of moisture. At the same time, these winds are rising higher and higher, as hot air is well known to do. In this way, the air passes through the regions of 20 degrees latitude without dropping moisture, and thus these areas become deserts. As the winds arrive over equatorial regions, they come into contact with colder upper regions of the atmosphere, cool down, and release all the water to help form the tropical rain forests near the equator.

Deserts also form in areas where a mountain barrier obstructs rain-bearing winds, and also areas such as in the center of large continental masses, where the source of water lies at a great distance. There are more desert areas in the world than most people realize.

Q. *Can the wind in a hurricane be strong enough for sticks, even a straw, to penetrate a plate of steel?*

A. It is possible, but the steel plate can't be several inches thick. The power of the wind can only be appreciated when a person experiences it. It is awesome. Even "moderate" winds of 70 to 80 mph can set people rolling in the streets. A straw can impale a glass plate during a hurricane. It is not unusual to see wooden fence posts in Texas embedded with thumbnail-sized pebbles. Some of the pebbles are three feet off the ground; mute testimony to the power of the wind. Sailors aboard the old Spanish galleons caught in Caribbean hurricanes reported that sea spray struck out the eyes of some of the crew. Though this seems like an exaggeration, we cannot be sure.

Q. *I know almost anything is possible these days, but is it true, as I've heard, that there are still people who believe Earth is flat?*

A. As incredible as it seems, there are several Flat Earth Societies whose members hold the belief that Earth is flat. Furthermore, their doctrine considers that continental drift is actually the land and oceans being shaken apart by God. They maintain that Earth is not a planet but a flat sheet. It is an infinite world, without end. Flat Earthers define themselves as seekers of the truth regarding the geophysical Earth. They consider theory as imaginary and instead focus on knowledge that is "provable." Some are almost viciously antiscience, calling science the "opium of the masses." Among their "proven" concepts are notions that the Sun is 3,000 miles from Earth, 32 miles in diameter, and the same size as the Moon. They believe Earth is 6,000 years old. They say it is logically impossible for people in Australia to be walking around with their heads hanging downward. Earth is also supposed to be stationary, neither moving through space nor spinning on its axis. Science does not, of course, support their claims. It is ironic that all the accoutrements of society that these individuals enjoy daily—among them the automobile, TV, radio, electric light and heat, medical advances—all come from the science they denigrate.

Q. *Where does all the sand we see on beaches come from?*

A. Streams inland from a shore are busy weathering and eroding rock material. This material slumps into a stream or river and is transported downslope, often for many miles. During this transport, size reduction of particles takes place due to abrasion and impact. A mixture of particles arrives at a beach on the ocean or a large lake. There, the waves' to-and-fro action sorts out the finer particles of silt and clay from the sand-sized particles, which are left on the beach and accumulate to form the shoreline. Finer particles are swept out into deeper waters. Under a microscope, these bits of sand are seen to be of different mineral origin, although most of it is quartz. In tropical areas, the sand may be made up of fragments of coral or other marine organisms that possessed a calcareous skeleton. Sand is a size term and does not indicate composition. Sand consists of particles between 2 mm and $\frac{1}{16}$ mm in diameter.

Q. *Why are some shorelines wide sandy beaches while others, such as the coast of Maine, are rocky cliffs?*

A. Geologists recognize both *submergent* and *emergent* shorelines. In the former case, sea level has risen relative to the land with deep water off the headlands. Large waves, especially in storms, strike the cliffs and cause erosion until a cliff is formed. In the case of emergent shorelines, the coast has been uplifted and the water offshore is shallow. Thus waves approaching the shore lose much of their energy because of friction with the bottom. These waves can nonetheless pile up, and sorted-out sand from finer sediment can be carried down to the coast by rivers. The beaches of the Carolinas are an example of this type of shoreline. In any case, shoreline erosion is responsible for millions in damage to property and recreational areas.

Q. *What is the deepest part of the ocean?*

A. The old idea that the floor of the ocean was a flat, featureless plain is largely untrue, although there are abyssal plains that fit that concept. The ocean bottom has varied topography. The deepest parts of the ocean are in narrow, slitlike *trenches*. Some of these trenches may extend 500 miles. The depth in one of these, the *Marianas*

Trench, was found to be 36,400 feet. The depth was determined by echo soundings and exploration in manned submersibles called *bathyscaphes*. It has been found that there are twenty major trenches, seventeen of them in the Pacific Ocean, and associated with considerable earthquake activity. It is at the juncture of plates and continental crust that subduction of plate material takes place. In the process, a trench forms as the physical expression of the subduction mechanism, which is highly complex and not yet fully understood.

Back in the nineteenth century, marine exploration was hampered by a lack of proper technology. Water depths were determined by a rope attached to cannon balls, which were quite inaccurate. Later, diving bells were used such as that of Charles William Beebe, an early marine explorer, but were limited in depth to the parting strength of the cable to which they were attached. Oddly enough, the deepest parts of the oceans—the trenches—are very close to land.

Q. *Giant tidal waves called* **tsunami** *can reach a height of 200 feet, they say. This seems unreal. How do they get close enough to measure it?*

A. We agree that 200 feet is unrealistic, but tsunami of 120 to 130 feet have been documented, and even tsunami of 50 feet or so are impressive enough. Some people imagine the tsunami as a wall of water sweeping in to the shore. The actual manifestation is a rapid rise of water to abnormal levels as it reaches a seacoast. We can estimate height in the aftermath by observing direct water damage above normal sea level. This would be along the same lines as viewing watermarks on your basement walls after flooding to see how high the water rose. The word *tsunami* has been borrowed from the Japanese who have had the opportunity to observe tsunami firsthand in their earthquake-prone country.

Unlike ordinary sea waves produced by wind action, the tsunami is produced by a severe blow to the ocean floor, either an earthquake or a volcanic explosion. Following such an event, the wave travels at several hundred miles per hour across open ocean, and builds up to great height as it enters the shallower coastal areas. While at sea, the tsunami is hardly noticeable, perhaps being only two or three feet high. The Pacific Ocean area is more likely to produce tsunami than elsewhere because of the earthquake activity there. Perhaps one or two occur each year. In 1946, tsunami hit the Hawaiian Islands, killing 159 people and causing $25 million in damage. The most destructive tsunami on record was the one that struck Awa, Japan, in 1703, killing an estimated 100,000 people. It should be noted that tsunami are not tidal waves, since they have nothing to do with tides.

Q. *I was looking at a bottom topography map of the Pacific and noticed numerous flat-topped mountains, appearing as if the tops were cut off. How are these peculiar mountains formed?*

A. These are volcanoes constructed of basalt and usually with gentle side slopes, called *seamounts*, that dot the Pacific and Indian Ocean floors. They are generated from hot spots lying 50 to 60 miles below the ocean bottom. About 2,000 of them have been mapped. They average about 3,000 feet in height. Why are they flat on top? It appears that most of them were at one time close enough to the surface to suffer erosion and truncation of their summits by vigorous wave action before subsiding to greater depths beyond the reach of the waves. Sediment samples recovered from the summits seem to bear this out as the summits are encrusted with coral formations. Corals like shallow water. The skeletal remains of these particular corals are of Late Cretaceous age, older than 60 million years. Another name applied to these features is *guyot*.

6

A FEW SKETCHES
Physics and Chemistry

Q. *Why is mathematics called the "queen of the sciences"?*

A. All scientists have a need at some time or other to measure things in their work and this requires at least a fundamental understanding of mathematics. Another explanation is that a pure mathematician need not know any other science to carry out his work, whereas physicists, chemists, geologists, and all others must know mathematics before they can work in their fields. It was the Babylonians who first established a science of mathematics, but the Greeks made strong contributions, especially Euclid, who wrote a book called *Elements* in 300 B.C.E., laying down the foundations of math and geometry. It is estimated that Euclid's book, in human history, is second only to the Bible in its circulation. While math is a tough struggle for some students, it is one of the most useful, even necessary, areas of learning. All of us use it every day.

Q. **With all the computers we have today, I sometimes wonder, who was the person who invented the computer?**

A. There is no single person who is responsible for the computer. Perhaps the forerunner was the simple counting device known as the *abacus*, which has been around for more than 2,000 years. Even before that, due to mathematical necessity, various crude computers or methods of computation were in use. For example, an army about to go into battle would be concerned about casualties. All soldiers would place one rock in a pile. After the battle, each surviving soldier would remove one rock. The remainder represented the losses.

As for modern electronic computers, the first was made in 1946 and called ENIAC. It was designed by J. P. Eckert and J. W. Mauchly of Philadelphia. It was a bit clumsy, containing 18,000 vacuum tubes. Later, UNIVAC was built, and twenty similar types followed within a few years. The whole field then expanded with better and better computers, as we know, until the present dynamic industry was reached. About 42 percent of households in the United States have a personal computer, and they have transformed the way business is conducted in the United States and increasingly around the world. The pace of advancement in the past twenty years has been spectacular. We are by no means at the apex of computer growth and development.

Q. **Where did our common measuring units of pints and quarts come from?**

A. The origins are probably in the England of the Middle Ages. Over the years there has been variation in what constitutes a pint, and also whether we are talking about liquid or dry measure. At that time, the quart was a dry measure while a pint was generally a liquid measure and equal in volume to a modern quart. You would need measures to sell something—-like beer or ale. So in medieval times, you would buy a pint of ale but get about a quart. Half of that was a cup for the conservative beer drinker.

Wine was measured in smaller quantities. A gill (or jill) was a standard measure for a glass of wine—-about four ounces or a half a cup. A jack was two ounces equivalent to a shot of whiskey. In order to raise sales taxes (yes, they had them then, too) Charles I decreed that the jack would be scaled down. Naturally, then, the jill would also be scaled down. This is the origin of the Jack and Jill nursery rhyme in which (the) Jack fell down . . . and (the) Jill came tumbling after. A ten-gallon hat holds only about a gallon; a cowboy would look ridiculous if it did hold ten gallons. This measure indicated the amount of water held and carried to the cowboy's horse over as many trips as the cowboy cared to make, or the horse to drink.

Q. *The speed of light is about 186,000 miles per second. How were scientists able to measure the velocity of something so fast?*

A. In 1676, a Danish astronomer named Olaus Roemer was observing eclipses of Jupiter's satellite Io, when Jupiter was at its nearest point to Earth. He cleverly deduced that six and a half months later, when another eclipse of that satellite would take place at Jupiter's farthest point from Earth, he could get an idea of how long it would take light to cross the diameter of Earth's orbit around the Sun. The delay in arrival time from that theoretically predicted was 22 minutes, which translated to about 180,000 miles per second for the speed of light traveling to reach Earth. Corrections were later made and other methods were also used here on Earth to determine the speed of light. In 1850, the Frenchman Armand Fizeau and others used two mirrors separated at some distance, one of which was rapidly rotating and the other stationary. Light from the rotating mirror traveled to the stationary mirror and back to the rotating mirror. The amount of rotation of the mirror during the beam of light's journey permitted determination of the speed of light. Although nothing can travel at the speed of light, some of the galaxies at the outermost limit of our detection are traveling close to 90 percent of that speed.

Q. *Has anyone succeeded in inventing a perpetual motion machine?*

A. Not that we know of. Over the centuries, many have tried to build such a machine, i.e., one that would keep working indefinitely, if not forever, without drawing upon any outside energy source. It has held as much fascination as did alchemy in the Middle Ages, which sought a way to change lead into gold. The most common type of perpetual motion machine was an overbalanced wheel; others relied on a continuous circulation of water. All such efforts fail because they violate the scientific law involving the conservation of energy (the total energy is constant). Yet there were many hoaxes and frauds. In 1870, John Keely raised a lot of money by claiming that a motor he had invented could power big ships across the Atlantic. Some devices such as a self-winding watch may appear to be a form of perpetual motion, but there is an outside energy source in the movements of the wearer. You can't fool Mother Nature.

A Few Sketches: Physics and Chemistry

Q. *Living in the atomic age, I still don't understand what a "chain reaction" is. Could you explain it?*

A. In 1942, during the Second World War, the first chain reaction was carried out at the University of Chicago. Physicists directed a stream of neutrons at atoms of uranium 235, causing them to fission or split. When that happened, the fissioned uranium nuclei in turn emitted neutrons which fissioned more uranium. It was a self-sustaining reaction, and produced energy. If such a reaction is allowed to accelerate quickly, as in a fraction of a second, then an atomic explosion will occur. If the chain reaction is controlled, then the energy can provide heat and electricity for peaceful purposes.

Several million years ago in Africa, long before man, a natural deposit of uranium went into a chain reaction and melted some of the surrounding rocks. Had this reaction been able to sustain itself and accelerate, an atomic explosion might have taken place on Earth without human witnesses—or victims.

Q. *Who invented the first thermometer?*

A. We believe that the first closed glass thermometer was invented by Cardinal Medici in Italy in 1654. He used red wine as the liquid because of its good visibility. Later in 1730 René-Antoine Reaumur used alcohol with a red dye. It was known that mercury would be a better liquid to use because of its more uniform rate of expansion with increasing temperature. However, the problem was to get a finer bore for the liquid to travel up and down and be easily observable. This problem was finally overcome. In 1724, Fahrenheit established 32 degrees as the freezing point, but 96 degrees (body temperature) as the high point rather than the 212 degrees for boiling water we are familiar with. Today there are thermometers to measure very high temperatures:

hydrogen thermometers can record up to 1,100 degrees Celsius and nitrogen thermometers up to 1,550 degrees Celsius.

Q. *Who invented neon lights?*

A. It was a Frenchman named Georges Claude. Neon gas had already been discovered by British physicists William Ramsey and Morris Travers in 1898. They had liquefied air and separated argon and then found another element that glowed when electrically stimulated. They called it *neon* after a Greek word meaning "new" or "recent."

Claude, who was a chemist, began experimenting with inert gases about 1910 and found that neon would glow orange-red when all electric current was passed through it, as had Ramsey and Travers, but Claude saw an immediate application to outdoor signs. The idea caught on quickly and as early as the 1920s neon signs became commonplace. This paved the way for fluorescent lamps and sodium lights. In turn, this helped changed our way of life in that we no longer had to stop working when it grew dark.

Claude made other important scientific contributions but got into trouble during World War II by collaborating with the Germans, for which he was imprisoned from 1945 to 1949. He died at ninety years of age in 1960.

Q. *Any schoolkid knows that Ben Franklin flew a kite in a thunderstorm and it had something to do with electricity. But just what was it he was trying to prove?*

A. We don't know for sure that this story is really true, and if it is, he probably flew his kite over in France. Franklin became very interested in all the sciences during the period 1748–1752. The purpose of the kite experiment was to demonstrate the electrical nature of lightning or *electrical fire* as it was called. Perhaps as a result of the kite experiment—if it really happened—Franklin proposed that buildings might be protected from being struck by lightning by the use of pointed iron rods. While most people think of Ben Franklin as a politician associated with the birth of our nation, he was a splendid and far-ranging scientist. He was interested not only in electricity, but also heat, sound, magnetism, geology, and chemistry to name a few. He invented many terms still used in discussing electricity such as *positive, negative, battery, conductor*, and others, and described the experiments clearly. Of course, he also invented bifocals. It would not be stretching it to call Franklin a genius. He was also pretty good with the ladies.

Q. *Does lightning never strike twice in the same place?*

A. This is popular folklore, but it is not true. Indeed, if a good place is struck by lightning once, there is a very good chance it will be struck again. High buildings such as the Empire State Building in New York have been struck by lightning many times. Once electricity has accumulated in the atmosphere, you need only positive and negative charge centers to cause the discharge of electricity in the form of lightning, which is really just a gigantic electric spark. This is not to say electric storms are neither awesome nor dangerous. Ancient peoples regarded thunder and lightning as examples of their god's anger or handiwork. In some African tribes, fires caused by lightning would be left to burn, and people struck by a bolt would not be helped for fear of interfering in the work of the god. Even today, superstitions persist. It is not uncommon for mothers to tell children that thunder is God moving furniture in heaven. Lightning may have played a major role in human advancement through the discovery of fire. Forest fires caused by lightning would have attracted human curiosity, leading eventually to the control of fire for warmth, protection, and cooking.

Q. *What causes the loud noise or boom when a plane exceeds the speed of sound?*

A. As long as a plane is flying at subsonic speeds, the disturbance to the air is well in front of the aircraft. As the plane builds up to Mach I (the speed of sound), a sharp pressure rise occurs in front of and tangential to the plane and creates a shock wave that manifests itself as claplike thunder. In a sense, the air molecules crowd together and collectively impact. It is interesting that the pilot of the plane does not hear the sonic boom, although on the ground, not only is it heard, but it can break windows. The effects of supersonic speeds were first described by the Austrian physicist Ernst Mach in 1881. Sound travels at a speed of about 1,100 feet per second at sea level, and this is why you might see someone, say, chopping wood at some distance, and see the axe strike before hearing the sound.

Q. *Why is ice so slippery?*

A. Ice has several unusual properties, one of them being that it melts when subjected to pressure. Your foot on ice is such pressure. When you apply pressure and ice begins to melt, a film of melted ice water develops, reducing the amount of friction. Thus sliding can occur. This is also why an object placed on ice can become embedded in the ice and work its way through the ice by melting ahead of and refreezing behind the object. However, small rivulets or streams flowing directly over ice cut down through the ice because the friction between flowing water and the ice generates enough heat to melt the ice regardless of pressure. About 99 percent of all the ice on the earth is found in Antarctica and Greenland.

Q. *I see the name Otis on so many elevators. Did he invent them?*

A. Elisha Graves Otis made the critically important contribution of safety features for an elevator so it would not fall. As early as 2,000 years ago, the Romans were using crude elevators to hoist freight or materials to higher levels during building construction. However, the ropes made of hemp were untrustworthy and often broke, sending the platform to the ground. A person would not wish to be a passenger on these contraptions.

Otis was a mechanic working for a bed factory in Albany, New York, prior to the Civil War. He was handy with tools and invented several labor-saving devices. He was promoted and sent to New York City. There he built an elevator equipped with clamps along the guide rails of the car such that they would grab and hold the elevator in place should the tension on the hoisting ropes be released. In a flamboyant demonstration in May 1854, he allowed himself to be hoisted high in the air and ordered the ropes cut. The safety device worked and the use of the elevator for passengers spread rapidly, permitting the construction of high-rise buildings. By the 1880s elevators were electrically driven, and push buttons were in use before the turn of the century. Otis helped change the urban landscape of the world.

Q. *When I see the Goodyear Blimp at football games, I wonder when the first blimp was invented.*

A. For more than 200 years, perhaps longer, man had tried to fly in balloons but was largely unsuccessful because there was no suitable engine to propel the airship. In 1851–1852, the Frenchman Henri Giffard built a ship equipped with a steam engine. It worked. He flew his 144-foot-long vessel over Paris at the breathtaking speed of 6 mph. Since that time, many dirigibles have been built in many countries. A few hot air balloons were deployed for observation during the Civil War (1861–65) and then again during World War I (1914–18).

Most of the early airships were filled with hydrogen, which has great lifting power but is highly flammable. The disaster causing the *Hindenburg* to burn in 1937 at Lakehurst, New Jersey, killed thirty-six people. There is still some dispute as to whether it was the hydrogen or the highly flammable outer skin of the air bag that originally ignited. Nevertheless, helium is the safer gas now used. By the way, the *Hindenburg* could cruise at 78 mph, a big improvement over Giffard's 6 mph. Goodyear has been a major builder of airships in the United States since 1911.

Q. *How far back can you date using tree rings?*

A. Further than you might think. Those old patriarchs, the California sequoias, turned out to be youngsters when it was discovered that the Bristlecone pines which grow near the timberline of western North America live to be over 4,000 years old. *Dendrochronologists*, as tree-ring scientists call themselves, have been able to build a tree-ring chronology by recording each annual ring in a living tree, then extending the count back in time by overlapping with rings in dead wood. By cross-dating rings this way, the Bristlecone chronology now goes back about 8,000 years.

Q. *Is there any possibility that science might some day build a time machine to travel into the past or future, or is it all fantasy?*

A. As intriguing as the idea is, we would have to say it is fantasy. The past involves events that had a beginning and an ending. It is a one-way street. Otherwise, we would have to allow that, say, Julius Caesar, here in the universe, must perpetually be having his diapers changed, cross the Rubicon again and again, and be forever stabbed by Brutus. It does not seem reasonable or even fair to Caesar. To travel into the future is to view something that does not even exist. However, one aspect to this fascinating topic is that we can view the past right now. We look out at stars many light-years distant. We see them not as they are now, but as they were when the light left them. For example, a star eight light-years distant is seen as it was eight years ago, not as it is today. An astronomer on a planet 2,000 light-years away, had he a magical telescope of superpower, might see Caesar cross the Rubicon. A mythical astronomer five billion light-years away might see Earth itself come into existence. In fact, we see galaxies forming at the edge of the universe in our telescopes now, but the event we are looking at happened billions of years ago. The present condition of such a body is unknown, since we are looking at past events.

Q. *Who was the greater genius, Thomas Edison or Albert Einstein?*

A. We would have to reach agreement first on what constitutes genius, which is not so easy. We suppose all geniuses are individuals who break new ground in the advancement of human knowledge and create whole new areas of science and art. Edison was in a sense a gadgeteer who transformed an idea into a working machine, but only after a great deal of trial and error. Einstein made no gadgets. His strength was theory—new ideas and concepts. For both of these men, genius was the creation of new foundations upon which other minds built and extended. Another way to look at genius is in terms of versatility. Leonardo da Vinci was a great artist, engineer, architect, and scientist. He also designed "gadgets" such as the parachute and submarine. Leonardo would be celebrated today were he to be remembered only for his contributions in one of these areas. Ben Franklin might be another, adept at science and invention, as well as politics and social organization (publication, the post office). The word *genius* is often used loosely to describe an extremely bright person. However, we hold the opinion that true genius is rare, and such persons as mentioned here come along all too infrequently.

Q. *The ancient alchemists tried to turn lead into gold. Did they ever succeed in anything important?*

A. Yes. They laid the foundation of the science of chemistry. In medieval times, there were many that called themselves alchemists but might be better called mystics or charlatans. The practical alchemists were for a long time obsessed with changing base metals such as lead into gold. They spoke of a philosophers' stone that would achieve this. Actually, the philosophers' stone might not have been a stone at all, but rather a substance or even a laboratory procedure. This serves to illustrate that alchemists wrote their experiments and books in cryptic, symbolic language, perhaps to guard "trade secrets" or in fear of being charged with heresy or blasphemy by ecclesiastical authorities, serious crimes in those days. Alchemists designed much of the laboratory equipment used today, such as beakers, flasks, retorts, furnaces, and distillation apparatus. They recognized and described several new classes of chemicals such as caustic alkalis. Gradually, alchemy merged into chemistry and became systematic. It is interesting that transmutation of one element to another, dreamed of by early alchemists, has been accomplished by modern nuclear physics, which has created several radioactive derivative elements by bombardment of unstable isotopes with subatomic particles.

Q. When did carbonated beverages come into vogue?

A. About 1807 in the United States, both Joseph Hawkins of Philadelphia and Benjamin Silliman of New Haven bottled and sold carbonated beverages. Prior to that time, people of infirm health visited spas to bathe in and drink mineral waters. Much mineral water is naturally fizzy. It was widely believed that such carbonated water possessed curative powers. This inspired scientists to seek a way to produce artificial mineral water. Joseph Priestly and others achieved this in the late nineteenth century. Basically, ordinary water is passed through an enclosure or chamber in which carbon dioxide gas is maintained under considerable pressure. The gas becomes dissolved in the water. The higher the pressure, the more carbonation takes place. Once the pressure is removed, as when one opens a bottle of soda pop, the carbon dioxide is released. This provides the fizz. We are familiar with the cola and fruity types of soft drink, but there are others. Soda pop can be made from various cheeses, and in Europe there is a carbonated beverage made from fermented stale bread.

Q. Why does water boil?

A. When you apply heat to a container of water, the heat energizes and excites the molecules of hydrogen and oxygen that make up the water. These molecules start crashing into each other. There are millions of these collisions, which release energy in the form of heat, and that is why the water gets hot, in the same way that if you beat a piece of metal with a hammer, the metal gets warm. As the agitated molecules bounce around, they tend to separate from one another and occupy a larger volume. When the density decreases enough, bubbles will rise through surrounding water that is denser. The result, of course, is steam. If the steam, as water vapor, continues to rise high into the sky, it will cool, condense, and regain its liquid form. If you are outside when this is happening, be sure to have an umbrella.

Q. *Why doesn't water have any calories?*

A. There is nothing in water that provides a source of energy for the body in the sense that something like carbohydrates can be "burned" by the body to create energy. Yet as we all know, without water we will not survive long. Water in our bodies allow, and participates in, the chemical reactions essential to life. Throughout history, people have recognized the magical qualities of water, or for that matter any liquid. Water is the universal detergent. Such ceremonies as baptism represent the cleansing of the soul in much the same way as water cleanses the body of dirt. Water is associated with life and the renewal of life. Thus, the Spanish explorer Ponce de Leon sought the Fountain of Youth in Florida and landed where St. Augustine is now. It is ironic that people in the twilight of their lives go to St. Augustine to await death in the same place he sought eternal youth.

Q. *How does a battery-powered flashlight work?*

A. The battery converts chemical energy into electrical energy by stripping atoms of their electrons and causing the electrons to flow, which is of course an electric current. The stream of electrons passes through the filament in the bulb and heats it, causing it to glow. The concave reflector in the flashlight directs the glow so you can see better. We know that three billion billion electrons flow through an ordinary 60-watt bulb every second. The metal tungsten makes an excellent filament for the electrons to pass through because it has the highest melting point of any metal. For example, at a certain temperature iron will boil while tungsten will still remain a solid. Thus, tungsten can resist the high temperatures generated.

Q. *How does mercury stick to glass when mirrors are made?*

A. It doesn't seem to stick to anything. In the old days, a flat sheet of tin foil would be laid out and mercury spread on it. The tin and the mercury would form an amalgam or alloy that enabled it to stick to glass. Then very carefully a sheet of glass would be lowered until it made contact with the mercury. Weights were then placed on the glass to squeeze out the excess mercury. Prior to this, the only mirrors were made of polished metal. In the Middle Ages, thin sheets of reflective metal were used as backing on glass. In the nineteenth century, Justus von Liebig, a German chemist, discovered how to make mirrors by silvering. This was better than using mercury. An ammoniacal solution of silver is used in conjunction with Rochelle salt or some other compound.

Q. *When did ink first come into use, and what is it composed of?*

A. Ink was being made and used for writing and drawing as long ago as 4500 years. At that time, it was used in both Egyptian and Chinese civilizations. They took lampblack, a pigment of nearly pure carbon, collected as soot, and mixed it with a gum or glue, and dried it out in sticks. When used, the sticks were swirled around in water to create the ink. Even today, ink consists of the same two basic components, a pigment and a liquid. Many different dyes and colored juices have been tried from various plants, animals, and minerals, but for the blueblack types of inks, iron (ferrous) sulfate is often used.

Q. *Why does some soap float, and how is soap manufactured?*

A. The soap that floats has small air bubbles in it, which make it lighter than water. The bubbles are retained in the soap during the manufacturing process by *aeration*, which involves rapid chilling of the soap material. Although soap chemistry is very complex, it involves basically a reaction between some kind of fat and a caustic solution (such as sodium hydroxide). This produces soap and glycerol. Part of the soap-manufacturing process is to separate the glycerol. Early Roman writers describe the first soap being made by boiling together goat tallow and wood ashes, so soap has been around a long time but not in mass quantities. Even as late as colonial times in America, each household made its own soap. With a general lack of soap until the twentieth century, people did not bathe much, and one account tells of a young woman taking a bath for the first time and finding out that it wasn't too bad once you were immersed and stopped shivering. Kids in those days must have been happy: No Saturday night bath. A final point: shaving cream is often mostly cold cream and not true soap, and the chief beard-softening ingredient is—water.

Q. *The use of cosmetics is such a big thing nowadays, what did women do before there was a cosmetics industry?*

A. Women—and often men—have been using cosmetics since civilization began. Probably the roots of the cosmetics industry are in ancient Egypt. There are murals 4,500 years old showing women with hairdressers using combs and polished metal mirrors. The tombs of that era contained urns and jars of aromatic oils and unguents. When some of these were unearthed after 4,000 years, the scent was still detectable. The Egyptians imported many of the ingredients from Arabia. It is known that women many centuries ago used perfume, eyeliner, and skin softeners. The women had the trick of placing the black substance *kohl* directly into the eye and then blinking to produce instant eyeliner! However, one thing they didn't have was the modern permanent wave. This was introduced about 1906 and was popular, even though a woman had to sit for ten to twelve hours, endure a fair amount of discomfort and pain, and pay a lot of money for it. The growth of the beauty shop industry came at about the same time because the home did not have the equipment needed.

Q. *How did the ancient Egyptians make papyrus?*

A. Actually, they made paper out of the papyrus plant. It grew in the delta of the Nile as a reed with a long slender stalk. The Egyptians sliced the stalk, or stem, lengthwise to form strips and placed them side by side. They then placed other such strips at right angles to form a sort of weave. After soaking in water, the weave was beaten with a mallet, dried, and smoothed—and you had a sheet of paper. Several sheets were joined together to form a roll. The person or persons responsible for this process are not known. The Egyptians used the same papyrus reed to make boats, sails, and even sandals. It was also edible. Shoes you could eat.

Q. *How is kerosene produced?*

A. It is one of the components obtained from crude oil. The oil, as it comes from the ground, is sent to a refinery where it is heated. As this is done the most volatile hydrocarbons (e.g., gasoline) are driven off first, leaving a residual fluid. After that comes kerosene and then increasingly thicker and less volatile oils and greases. When drilling for oil more than a century ago, it was to obtain the kerosene for illumination in lamps. This replaced whale oil and society entered the "kerosene age." The gasoline, which had to be drawn off first, was troublesome and dangerous. Early drillers dumped it in creeks where it caught fire frequently. At the turn of the century, with the development of the internal combustion engine, a use was finally found for gasoline and we entered the "gasoline age." Even then, many people said gasoline use in automobiles would cause them to blow up.

Q. *Is it true that the Chinese discovered gunpowder?*

A. Maybe not. The English, Arabs, Hindus, and Greeks say they did. We know that the ancient Greeks concocted *Greek fire*—a sticky mixture of sulfur, pitch, and oil that terrorized their enemies when thrown at them, not only because it stuck, but its flames were hard to put out. Perhaps this was the beginning of gunpowder, or "black powder," as it was otherwise known. Gunpowder is mainly saltpeter, or potassium nitrate, which burns fiercely when blended with about equal amounts of charcoal and sulfur. If confined in a container of sorts, there is an explosion. Perhaps in this way 600 to 700 years ago gunpowder became something to use to blow things up. Some historians think the advent of gunpowder signaled the end of the large castles of the Middle Ages because the walls could be blown apart. Without a doubt, gunpowder changed not only warfare, but also human history.

Q. *What is the difference between gunpowder and dynamite?*

A. Gunpowder is a low-order explosive and safer to handle. It was used early on as a propellant for guns and for blasting in mining operations. When ignited within a confined space, the large amount of gas produced serves as the propellant. Today, it still has wide usage in fireworks, military practice bombs, or other areas where a lot of smoke is desirable, but without the danger of a truly devastating explosion.

The key ingredient in dynamite is nitroglycerin, which by itself is unstable in handling and produces a violent explosion. Alfred Nobel, in experimenting with nitroglycerin, found a way in 1867 to control its explosivity by admixing it with siliceous earth. Then it was found wood pulp would work as well or better. This resulted in a safer, wider use for dynamite, thereby making a fortune for Nobel who used part of his money to establish the Nobel Prize.

Nobel was a unique man. He came from a large family and his father was in the explosives business. Nobel became a chemist and had several inventions and patents to his credit in detonators and explosive types. This work, one may imagine, was not without danger. At one time Nobel's factory blew up, killing his brother Emil among others. There were other such accidents during his work with explosives. Yet due to his work, numerous construction projects for bridges, tunnels, highways, foundations, and other purposes were facilitated.

Q. *Why does science dispute the Bible?*

A. Science does not dispute the Bible except where it is claimed that the Bible represents scientific truth. The Bible is a source of religious faith and does not require the exercise of reason for its believers to follow its teachings. In that context, science is not involved and does not care to get involved. It is only when enthusiasts of the Bible claim there is scientific evidence to support biblical assertions that science enters the picture. Each claim should be judged on its own merits. The cosmology of the creation depicted in the Bible runs counter to all the evidence amassed over centuries by hundreds of scientists in many lands. The statement that the worldwide flood of Noah as described in the Old Testament is supported by geological evidence is not true. Quite the contrary. Yet other aspects of the Bible may have some historical roots, subject to the data and testing of science. It should be noted that many scientists accept the Bible as part of their religious life—a source of inspiration, but not a scientific textbook.

PART 3

LIFE
ON
EARTH

7

GENES, SURVIVAL, AND EXTINCTION

Q. *Exactly how does one define and describe evolution as scientists understand it?*

A. Evolution is an explanation of how all life on Earth originated, developed, and diversified from preexisting forms, and the natural mechanisms, both internal and external, that brought this about.

In the mid-nineteenth century, Darwin proposed that new species arose by a process called *natural selection*. The variation in offspring of a species would yield individuals possessing either favorable or unfavorable traits germane to survival within an environment. Those with favorable traits would live, reproduce, and pass on the favorable traits to their offspring. The others would die off. It was a process of selective pruning. In time, this would result in a new and separate species. Mendel showed later that these traits were inherited in definite proportions. By 1920, it was known that puzzling deviations from Mendel's Law were due to mutations of genes and chromosomes. Mutations could be adverse, neutral, or in a minority of cases, a leap forward in the evolutionary process.

The 1950s saw the growth of molecular biology and an understanding of DNA, which represents the code or blueprint governing details of an organism's physical development. This DNA is found in every cell in the form of nucleic acid. The continuing revelations of molecular biology demonstrate exactly how evolution has functioned throughout geologic history. In a sense, molecular biology is the grammar of evolutionary language. The final proof of evolution resides in the fact that humankind is now on the threshold of controlling and manipulating evolution. Increasing experimentation of cloning is a step in this direction.

Q. *I hear a lot about genes and chromosomes. What is the difference?*

A. Genes are tiny carriers of hereditary traits that through interaction with other genes produce the physical and other characteristics that determine the nature of an organism's offspring. Chromosomes are collections of genes and are threadlike. Humans have twenty-three pairs of these chromosomes and thousands of various genes. We might draw the rough comparison that the chromosome is like a suitcase and the genes are the contents of the suitcase. In sexual reproduction, the parents contribute equally a packet of genes and chromosomes and in the process of mixing, so to speak, there are almost limitless combinations possible due to sexual variation. This is why everybody is different, even offspring of the same parents, and is a key factor in the process of organic evolution. In the future, it may be possible to remove undesirable genes and replace them with favorable genes and so control the development of offspring. This will entail some ethical questions because not everybody may agree on what is undesirable or favorable.

Q. *How can one species of animal or plant develop into two species?*

A. The original species may have members that become isolated from the rest of its kind. There could be a physical isolation into two subenvironments. For example, during a storm at sea a few individuals of the same species are thrown into a lagoon by a large wave. They inbreed and adapt to the quieter, warm, and shallow waters of the lagoon. With time they become so different from the original species that they are unable to interbreed with them. The same thing could happen if one group stopped interbreeding with the rest through sexual preference. Thus, the genes would not mix even though all members of the species remain in the same environment. This *genetic drift* would ultimately result in a new species. When two groups can't interbreed, they are classified as separate species.

Q. *I am amazed at the delicate balance between living things and the environment. For instance, if the planet's temperature were just a little hotter or colder, so much life would be ended. Has science an explanation?*

A. We think so. Consider the offspring of a particular species. They are not all carbon copies. There is variation in traits due to the differing assemblage of genes each individual received upon conception. Some individuals will not be able to cope with the environment if, for example, the food supply or climate is adverse. These individuals will die. The hardier individuals will survive and pass on their traits to their offspring. Thus, a species may become better and better specialized in dealing with its environment. Scientists consider this process a *gene pool shift*, important in the development of a new species, and one that is well adapted to all aspects of the environment. And so the creatures of Earth achieve this "delicate balance" with their surroundings, but it doesn't happen overnight. More often than not, millions of years are required. There is a danger, though, in that should a radical change in the environment occur quickly, an overspecialized species may not have time enough to adapt, and this could lead to extinction.

Q. *Can you clarify the debate going on between evolution and creationism?*

A. In science classes, evolution is taught as a science because it is one. Since Darwin first proposed evolution, there have been arguments among scientists and changes of opinion as new findings were reported. The enormous body of information gathered from such varied disciplines as paleontology, biology, medicine, comparative anatomy, genetics, serology, and even man's own breeding experiments has solidified the fact that evolution did occur and is taking place now. It is not "just" a theory. The lifeblood of science is an attitude—one of gathering data, testing ideas, and changing them as the findings dictate. Science is mutable almost by definition. Evolution is a case in point.

On the other hand, creationism is a belief in how life came to be, and is based upon religious scriptures which, since they are perceived by many to be the word of God, are not subject to change, testing, or debate. That is not science. The difference is in their methods and attitude. One is science, the other is not, yet both points of view should be accessible to inquiring minds. That is education. The appropriate place for creationism to be discussed is in a cultural or religious context. To present creationism as science in any school is no more logical than teaching oil painting in a physics class.

Q. *Why do scientists say we evolved from a monkey?*

A. What is interesting about your question is that this idea was probably voiced by someone unschooled in the subject, but opposed to the concept of evolution, stating what they thought scientists were saying. It is a misconception. Scientists hold nothing of the kind. Humans did not evolve from the monkey and may not even have evolved from the ape. Certainly we are more closely related to a monkey than we are to, say, a chicken. But at best we are only cousins. Scientists assert that humans, monkeys, and apes evolved from a common ancestor several million years ago. The data supporting this conclusion is not as skimpy and speculative as some make it out to be. But scientists are notorious for playing down the information that they have. This is why even today scientists refer to evolution as *theory* when in reality, that evolution has occurred is hard fact, based upon many lines of evidence and is the foundation stone of modern biology.

Q. *Is the giraffe's long neck the result of stretching for leaves on high trees?*

A. No. In the nineteenth century, some scientists such as Jean-Baptiste Lamarck thought that acquired characteristics of an individual could be passed on to its offspring. Likewise, traits not used much would gradually disappear in the offspring. This was called the *law of use and disuse*, and is no longer held to be true today. If it were true, then a weightlifter (for example) would pass on his built-up muscles to his children, which we know doesn't happen. Traits such as the giraffe's long neck are inherited via the genes. Where genetic characteristics represent an environmental advantage to the organism, the latter will survive and pass on these successful traits. Traits that do not equip an organism for survival within its environment may doom it to extinction.

Humans walk upright while other primates don't. Why is this said to be a big "evolutionary advance"? Try getting down on all fours and driving a nail with a hammer at a place above your head. You can't. If the forelimbs and hands are involved in support, this will severely restrict their use. However, standing on two legs frees the hands and arms for other tasks such as toolmaking, especially when the hands are equipped with opposable thumbs. It is this ability to make tools (and ultimately tools to make other tools) that accelerated humans' evolutionary progress to its present degree. Another important factor is that standing permits senses of sight and hearing to be elevated higher, conferring a big advantage in surveying the surroundings—a big plus for survival. If you add the effects of upright posture, freed hands, and large brains, the result is a creature capable of making not only tools, but also cave paintings, telescopes, computers, and spaceships.

Q. *Which are more intelligent, chimpanzees or gorillas?*

A. Probably the popular notion is that the chimpanzee has the higher intelligence, but this has not been proved. Both creatures are apes and closely related to humans. Indeed, some chimps can perform as well as some young human children in memory tests and spatial visualization, when language skill is not involved. The chimpanzee adjusts well to captivity and can be described as a very outgoing, extroverted creature.

In contrast, gorillas are shy and retiring creatures and so, not much is known about them. The first gorilla wasn't born in captivity until 1956. They like to eat and sleep most of the time, and are not the ferocious, deadly beasts depicted in books and movies. It may be that they are as "smart" as the chimp. Neither can match the human in intelligence because the human develops ideas and concepts to use in solving problems. Moreover, the human has speech and writing to hand down accumulated knowledge to future generations.

Q. *Is it true that scientists are beginning to find that animals use tools as we humans do?*

A. Yes, to a certain extent animals will utilize tools in their quest for survival. For example, the heron may use one of its own feathers as a lure on the water's surface. When a fish investigates the feather, the heron quickly scoops up its next meal. Monkeys insert a stick into anthills and allow the ants to crawl onto the stick. The monkey then withdraws the stick and licks off the ants. Also, simians may use clubs to drive off predators. Although clever, these are instances of casual use of available tools. One of the human's distinguishing characteristics is the ability not only to visualize and fashion tools, but also to use tools to make other tools. We don't know of other creatures that can do this.

Q. *Did Charles Darwin believe in God? His idea of evolution opposes so much religious thought.*

A. Darwin died before our time, but our reading of his life and work suggests strongly that he was a shy, modest, and deeply moral person. His life's work, leading to the concept of evolution in his role of biologist-geologist, actually produced within him considerable confusion about design in the universe, as he was reared according to the orthodox religious views of the Church of England. Perhaps toward the end of his life he could be described as an agnostic. In some religions, of course, evolution is regarded as anti-God, but not in all. The Catholic Church accepts evolution as God's method of creation, the concept being that the human body evolved as science claims but that at a certain point a soul was instilled into the human apart from evolution.

Darwin was not an armchair theorist. As naturalist aboard the *Beagle* he was able to study and observe animal and plant life all over the world during the ship's five-year voyage (1831–1836). He spent the next twenty years assimilating and organizing his data, which resulted in his seminal work *Origin of Species*. The wide interest sparked by his theory of evolution is reflected in the fact that his book sold out the first day of its publication and has gone through many editions. The regard of his countrymen is also shown in that, upon his death, he was buried next to Sir Isaac Newton.

Q. *How do insects become immune to insecticides?*

A. It is not that an individual insect becomes immune. Most insects will die right away, but a few will have a built-in immunity, will survive, and produce offspring that inherit this immunity to the particular insecticide to which they were exposed. This type of reaction is common to all living organisms and is a fine example of evolution in process. Another example is the invasion of New York City by "super-rats." Some years ago the city made a determined effort to poison all the rats in town, and at first the number of rats declined dramatically. But there were a few that withstood the poison. They have now repopulated their kind and the problem is no better.

Q. *When did creatures like birds and bees start flying and how did flight evolve?*

A. There were actually two evolutionary routes to powered flight in animals. The insects were first. There are winged insect fossils 300 million years old from the Mississippian period. Birds came much later in attaining powered flight, about 140 million years ago. The first bird we can recognize is called *Archaeopteryx*, which was about the size of a quail and possessed both teeth and feathers. The postcranial skeleton is clearly reptilian, and so we have a very fine not-so-missing link between birds and reptiles, in this case, dinosaurs.

Birds achieved flight by gradual modification of the forelimbs from walking to flapping. The bat is another example. We might speculate that proto-avis was a four-legged creature trying to catch a flying food supply—the insect. It did this first by leaping, then by gliding, and then by attaining flight in the quest for food. We do not know the evolutionary mechanism that brought about insect flight. Intermediate insect fossils showing this progression have not yet been found. We can, however, be quite sure that insect and bird/mammal wing structures are completely different.

Q. *What is cloning?*

A. Cloning is nothing more than a method of reproduction. If a human couple produces a child, the child would not be a clone because there has been an interchange of genetic material between the parents. If a simple cell divides into two cells (asexual reproduction), that would be cloning, because the genetic material is identical and without contribution of genes from outside. Experiments have been attempted since the 1950s to produce replicas or clones of lower animals with some success with sheep in Scotland and elsewhere. Whether or not human clones can be produced is a moot question. Right now adult cells cannot be cloned, so the closest duplicate of yourself you will see is probably in a mirror. If technology permits duplication of humans some day, it will raise a number of very interesting moral and legal questions.

Q. *What value is there in cloning animals and plants?*

A. It is a process of duplication. As you may know, in nature there are organisms that reproduce asexually; in other words, an individual produces one or more other individuals without combining or exchanging genetic material. Thus, the offspring are carbon copies. In higher organisms such as humans, there are always unique combinations of genetic material by sexual reproduction so that each of us is different in some way. Cloning allows the birth of new individuals without the addition of other genetic material. Research in cloning can be beneficial to humankind in providing a source of insulin, growth hormones, or other useful substances. Advances can also be made toward the cure of disease and an increase in food supply. The science fiction notion that you might walk down the street and bump into your exact double is a bit far out.

Q. *Isn't the "population explosion" just a myth?*

A. Yes and no. The idea goes back to Thomas Malthus, an economist who lived from 1766 to 1834. He said that population will increase geometrically while food supply increases arithmetically, meaning that the population would outstrip the food supply and people would die of starvation. To illustrate the meaning of this, start with one person and double the number each year for thirty years and you will have more than one billion people. In industrialized societies such as the United States, this has not held true because economic output has tended to lead population growth, so nobody starves. This is also due to widespread birth control.

However, in nonindustrialized countries (Third World), Malthus's theory holds generally true and people are indeed starving because the population growth is unchecked. If the human race cannot find a balance between food supply and population growth, the Four Horsemen—death, disease, war, and pestilence—will ride again, and they are already manifest in several African and Asian countries.

Q. *Why will the population explosion necessarily cause war, disease, and the like?*

A. It is a well-established biological fact that animals such as birds, dogs, and cats will "stake out" a territory that can't be invaded without a fight. As long as territories are broad, no fights will occur. However, compression of territory by multitudes of individuals leads to quarrels. It is instructive to consider a biological experiment in which white rats were provided with adequate space and food and allowed to reproduce with no restriction. At first, there were no problems—all rats had food and space. However, once the laboratory facility became overcrowded, fights resulted, and rat killed rat. In such conditions, disease is more communicable. This may be the reason why there is more crime in overcrowded urban areas. We really ought to spread ourselves out. About 98 percent of the United States population is concentrated in only 2 percent of the land area. Most of the country is empty. A better distribution would be helpful for all. However, humans are social creatures, herd animals if you like. That is why we have densely crowded cities. It is a real danger to our survivability.

Q. *If life began all over again on Earth, would it evolve to be the same as today?*

A. Probably not. Even though environmental conditions were the same in all respects, small accidental occurrences (mutation?) may well deflect the path of evolution to favor one group of creatures over another. The early mammals that started out nearly 200 million years ago, and eventually led to humans, were small and weak. Somehow they managed to survive through the Mesozoic era when reptiles, especially the giant dinosaurs, dominated Earth. There would be no guarantee that this would happen twice. Similarly, if two basketball teams played each other twice during the same season, no one would expect the final score to be the same in each case even though the players and playing conditions were the same. In another go-around of life evolving on Earth, the leading form of intelligent life on Earth might well be some insect.

8

ANIMALS
Sea, Land, and Air

Q. *Okay, which came first, the chicken or the egg? I have never seen an answer to this age-old question.*

A. You could also ask: which came first, the turtle or the egg, or the alligator or the egg, or any of the creatures that lay eggs. We don't think the answer that tricky if you take into account evolution. All creatures on Earth, past and present, represent life-forms that did not previously exist but yet had to have evolved from previously existing types. In the case of the chicken, the ancestors evolved over millions of year with successive generations becoming more and more chickenlike. Finally, the most chickenlike of the chickenlike predecessors hatched an egg. It was a true chicken. So the obvious answer is that the egg came first, but it required a very close relative to oversee the nativity.

Q. *Are bacteria considered plants or animals and will science be able to wipe out all bacteria to prevent disease?*

A. These microscopic creatures were at first thought to be plants of a sort, but not belonging to the animal kingdom. Nowadays they are relegated to a third kingdom, the *Monera*. There is a bit of waffling here, meaning scientists aren't yet certain of their classification.

We fervently hope science will not wipe them all out, otherwise we will all be dead. Many bacteria are beneficial and even vital to our survival. Consider that nitrogen-fixing bacteria keep plants going. Cows and other grass-eating animals could not digest plant material without bacteria in their stomachs. Bacteria also are our best waste-disposal allies. They break down dead organisms so as to recycle nitrogen, phosphorus, and carbon, without which we could not survive. While it is true that certain bacteria can cause disease, other more friendly bacteria assure the continuation, through their life processes, of some foods we may like such as pickles, buttermilk, yogurt, cheese, and even sauerkraut.

Q. *I read that whales feed mostly on plankton. Just what are plankton?*

A. The word itself refers to small organisms floating in the sea that are more or less at the mercy of the currents. Opposed to this are the *nektonic* forms with swimming ability. Plankton consist of floating animals and plants (*zooplankton* and *phytoplankton*) at the bottom of the food chain. Larger fish and other organisms eat these small creatures and are themselves eaten by larger creatures, and so on up the food chain to humans. Some plankton are the size of a small grain of sand; others are even smaller and will pass through the finest nets.

Plankton are of critical importance. Phytoplankton generate the oxygen we breathe. Scientists believe that some day these plankton will become the chief food source for the people of this planet. This question is interesting in that the whale, among the largest creatures on Earth, survives and is nourished by some of the planet's smallest inhabitants. Although the whale is an air-breathing mammal with a land ancestry, it could no longer live on land. Its weight alone would give it respiratory failure.

Q. *Do fish sleep? They seem to be moving around all the time.*

A. It isn't known if they sleep in the same sense that we do, but they do seem to have regular rest periods. Fish may not appear to sleep because, lacking eyelids, their eyes are always open. Some fish merely remain more or less motionless in the water, while others rest directly on the bottom, even turning over on their side. Some species are observed to dig or burrow into bottom sediments to make a bed of sorts. Some fish seem to prefer privacy when they rest, because their schools disperse at night to rest and then reassemble in the morning.

Q. *Can piranha actually reduce a cow to a skeleton in seconds?*

A. This is perhaps a bit exaggerated, but the cow involved may argue the point. Schools of these ferocious, freshwater fish, found in South America, have been known to attack a 100-pound capybara (a large South American rodent) and reduce it to a skeleton in one minute. The teeth are triangular, razor-sharp, and efficient even though each piranha is only a foot long. Piranhas have also attacked humans and although we cannot cite specific cases, probably with fatal results. Hobbyists have kept piranha as pets, but there is a danger that they could be released into streams and create serious ecological problems. Also among dangerous fish is the lion fish, which is found in the Pacific Ocean. They have poisonous spines on their fins with which they are capable of stabbing a person, which causes a painful but not lethal injury. The lion fish is about one foot in length and attractively striped.

Q. *Are stingrays that live on the seabed classified as fish? Can they harm humans?*

A. There are quite a variety of stingrays, or "rays," in general belonging to the same group as the shark (the *Chondrichthyes*). This means that they have cartilage rather than bone as the major part of their skeleton. Nonetheless they are fish. The favored habitat is warm, tropical water. Some species are able to deliver an electric shock while other species have a sharp, barbed tail that is poisonous. We would assume that pursuing these creatures on the sea bottom could be dangerous, since they will naturally defend themselves. A well-placed stab of the tail can be fatal. Yet they are fished for on a commercial basis, especially in Asia and Europe, and the meat is very good. The larger species (e.g., the devil rays) are 20 feet across and weigh up to 500 pounds. Usually they lie on the seabed, well camouflaged and waiting for food such as other fish and crustaceans. When they swim, they resemble large bats flapping along. Stingrays are responsible for the destruction of some commercial shellfish beds.

Q. *Can an electric eel electrocute a person?*

A. Actually, the electric eel is not an eel but a freshwater fish more closely related to the knifefish of aquarium hobbyists. Electric eels can exceed a length of 3 feet, and some reach up to 9 feet in length with a thickness of 8 to 10 inches. The body of the fish constitutes an electric storage battery in which the head and tail are negative and positive poles, which in contact with the victim will deliver a shock. The wider the spacing between head and tail in contact with the victim, the greater the shock. A charge of 650 volts has been recorded for brief contacts. Such magnitudes are sufficient to kill an animal even as large as a horse. Humans have been attacked and severely stunned, but we know of no deaths. There are other fish such as the stingray that have similar electrical capabilities.

Taking a broader view, electricity, or, better, *bioelectricity*, is involved in the normal functions of organisms. It takes the form of a flow of ions—charged atoms—across the membrane of a cell, which contains a different solution than the surrounding fluid. This produces a current. Messages sent along our nerve pathways are electrical in nature.

Q. *I have heard that the male seahorse gives birth rather than the female. Is this a biological fact?*

A. This is actually a myth. What must be distinguished here is the difference between reproduction and nurture. The female seahorse is the one that produces the eggs. These are then placed in a pouch near the tail of the male until birth occurs. The male is actually participating in the nurturing of the young, although at a very early stage. Note that in many species of creatures, for example, some birds, the male sits on the nest for periods before the eggs are hatched. This gives the mother the chance to go seek food. The seahorse is actually a fish and a member of the pipefish family, despite its unfishlike appearance.

Q. *What fish is most often caught and eaten, worldwide?*

A. Probably the herring, which is abundant in both the Atlantic and Pacific Oceans and forms an important part of the marine food chain. Herrings are found in schools by the thousands and caught in nets. A female may lay 40,000 eggs at a time. It takes a herring four years to reach adulthood and it might live to one hundred years if it escapes capture. A good-sized herring is about a foot in length. Besides being salted, pickled, creamed, and smoked, inedible parts of the herring are ground up to feed livestock. A major by-product is oil, and in Europe they make margarine from it. The canned sardine is a member of the herring family.

There is considerable folklore concerning the herring. For example, a man who eats a raw herring will see a true vision of his future wife. To brag about the size of your catch is to bring bad luck. It was believed that a school was led by a giant herring called the Royal Herring, and this fish should not be caught or harmed.

Q. *Is a barracuda as dangerous as a shark?*

A. Some attacks on swimmers that were thought to be by sharks were in fact attacks by barracuda. However, it should be remembered that the normal diet of the barracuda is other, smaller fish such as anchovies, and not people. A barracuda is long and slender with a large mouth and usually grows to lengths of four to six feet. Extremely swift fish, they have razor-sharp teeth and attack anything that moves. They are curious and bold which perhaps accounts for their unpredictable behavior. Some barracuda actually control herds of other fish in shallow water, waiting until they become hungry to eat them. Fishermen like to fish for barracuda because they are hooked easily, fight like tigers, and are a tasty item on the menu.

Q. *Can sharks smell human blood in the water from a mile away?*

A. One would have to assume that traces of blood could circulate to distances of a mile in order for the shark to have the opportunity to detect it. This is highly unlikely unless you are bleeding like a stuck pig. It is known that people have been attacked by sharks in shallow water and have suffered severe wounds and loss of blood without the shark resuming the attack. In some experiments, sharks swam through human blood without getting excited at all. They had the same response to rat blood. However, when dead rats were soaked with fish blood they were immediately attacked by the sharks. Humans are not a normal part of the shark's diet and so we do not command particular attention. Actually, sharks are more attracted to vibrations and movement in the water near the surface where disabled or dying fish might be located. Perhaps this accounts for the many shark attacks on humans. It may be a case of mistaken identity.

The vast number of television features, movies, stories, and articles about sharks indicate the public's fascination with the subject. Although the tiger shark, sand shark, and others are dangers to human bathers, the white shark is most dreaded because of its aggressive behavior and size (greater than 7,000 pounds recorded and lengths up to 36 feet). While those who saw the movie *Jaws* might think the shark attack on the boat another case of movieland sensationalism, sharks do indeed attack boats, even leaving behind teeth embedded in the wood. The American Institute of Biological Sciences established a Shark Research Panel to gather data on shark attacks around the world. They found that for the thirty-five years from 1928 to 1962, there were 670 attacks on persons and 102 on boats.

Q. *How big can an octopus get, or a squid?*

A. A respectably large octopus can attain a size of 35 feet, measured as arm-span, and a squid can be even larger at 60 to 65 feet. These are exceptional. Many members of these species, known as *cephalopods*, can be measured in inches, and provide a food source for many marine organisms. Along with their close cousin the cuttlefish, these animals enjoy a reputation as the most intelligent of the invertebrates. Predators themselves, they subsist on crabs, lobsters, and other crustaceans, and can be considered as bottom-dwellers despite their mobility. Yet on occasion they may be seen at the surface, giving rise to the many legends of sea serpents found in early nautical literature.

The squid is an ideal survival machine with large, excellent eyes for spotting prey; a jet siphon system for swift approach; ten tentacles with suckers for grasping prey; and a mouth equipped with sharp beaks and a rasping device (the *radula*). On defense, the squid can change color like a chameleon, retreat quickly with its siphon, and leave an inky smoke screen to confuse a pursuer. In some species, the ink is poisonous. It is little wonder that these remarkable creatures have survived in the seas for 500 million years (since the Cambrian).

Q. *Hasn't Hollywood exaggerated the danger of big snakes, tigers, spiders, and other jungle creatures?*

A. From our experience, we would agree with you. There is the tendency to magnify the danger presented by these creatures for the sake of drama and excitement. Most of nature's creatures simply want to survive and mind their own business without fighting with others. We have walked within five or six feet of boa constrictors without so much as a ho hum. Lions will lounge within a short distance of zebra herds without attacking if the lions are simply not hungry. Disturb a spiderweb and the spider usually flees. While these creatures may present a menace under certain circumstances, they are not what you would call "mean." Man's greatest enemy in the jungle is perhaps the mosquito or other insects, while at home the worst enemy might be ants in the kitchen. Some people have been driven to the edge of madness by prolonged exposure to mosquitoes. In tropical jungles, a person's hat can be covered by mosquitoes so thick that it resembles brown fur. In any case, one might feel safer in the densest jungle than in New York's Times Square at rush hour.

Q. *Why have alligators and crocodiles been able to survive?*

A. There is a bit of a mystery here because the mother is an indifferent parent. After laying twenty to seventy eggs, she guards the nest only on a part-time basis. It is known that small mammals such as raccoons will eat the eggs of the American alligator. At the point of hatching, the female will assist the little alligators into the water. Perhaps the major reason the crocodilia family has survived is that once hatched, the young are perfectly able to take care of themselves without nurturing. The greatest enemy of these beasts are humans who hunt the adults for their hides to make luggage. This alone has brought some species to near extinction. Despite the gaping mouth and numerous teeth, they can't chew their food, being forced to swallow it in big chunks. The crocodilia are closely related to the dinosaurs as both groups belong to the subclass *Archosauria*. Somehow, they escaped the fate of the dinosaurs, but for how long?

Q. *Could human activity bring about pollution of the oceans?*

A. If we broaden the definition of pollution to include all human activity harmful to oceanic life, then it already has. The world's population is concentrated along shorelines. We build our cities there. Our rivers have become huge sewer lines conveying municipal and industrial waste into shallow coastal areas that have become gigantic garbage dumps. These same coastal areas are the spawning and living areas for the bulk of oceanic life. Ten percent of the world's coral reefs are already destroyed, and more will perish during the course of the twenty-first century. Besides this, we have drastically overexploited food-fish banks that once teemed with life (such as cod and flounder). Moreover, oil pollution continues to harm aquatic biota. Water ballast in ocean shipping transfers hundreds of species from one marine environment to another, often with the unforeseen destruction of native populations. Human hunters continue to reduce whale and shark populations despite governmental decrees.

Of sinister import is the threat to the ocean's plankton, source of 70 percent of the oxygen we breathe. Red tides, or deleterious red algae, and other algal blooms linked to human activity are affecting plankton adversely. This, coupled with increased ultraviolet radiation due to ozone depletion, may well cripple the fertility of plankton. At present, there is no pat answer to halting the degradation of the oceans although our own survival depends upon them.

Q. *I was told that there are earthworms 11 feet long. Isn't this a tall story?*

A. It would only be a tall story if you were told somebody used this worm as bait to catch a fish. Yes, there is a species of earthworm 11 feet long found in Australia, and is the extreme in size among the 1,800 species of worms found in the world. There are also green earthworms in Great Britain although most are reddish-brown. Worms are blind and deaf but they respond to vibrations. Most worms burrow close to the surface but some can tunnel down to seven feet. Worms are hermaphrodites—possessing both male and female reproductive organs—yet they still need another worm for mutual fertilization. The chief function of the worm for humankind is to turn over the soil on which crops are grown. Fishing enthusiasts may disagree.

Q. *Can army ants actually bring down and devour a horse or even a person?*

A. No. Although army ants may ingest larger carcasses, they will not attack and eat a living person despite what movie studios may depict. Instead, army ants feed upon cockroaches, beetles, locusts, spiders, and other insects. And they are vicious. Any of these unfortunate insect victims can be covered instantly by a dense mass of ants and their bodies torn apart. In seconds, limbs and other pieces of the insect are severed by sharp pincers and moved to the rear of the ant army. Usually army ants scour a forest, especially tropical forests, in columns. Upon finding a target, the column divides into two directions to cut off escape. The ants resemble a real army, even foraging with advance scouts ahead of the column, attacking an area and abandoning it as its resources are devoured. Army ants will not attack other ant species except for the cowardly *Hypoclinea*, which run away, and even then only attack the larvae and pupae.

Some ants raid other nests and capture slaves. There are five ant species that do this. The belligerent queen of the raiding group and her consorts force their way into another ant group's nest. If worker ants resist the invaders, the queen quickly kills them. There are two ways in which the vanquished ants become slaves: (1) The discombobulated ants stop fighting the victorious queen and attempt to guard the brood of larvae. When the larvae grow into ants, they recognize the foreign queen as their own. The original queen will either have died of neglect or been killed during the invasion. (2) The invading ants gather up the pupae and larvae and take them back to their own nest where they develop and work for the other ant group.

Q. *How long do spiders live and are most poisonous?*

A. Most spiders seem to live about one year, but they can live longer in warmer climates. The large bird spider, as big as a fist, lives longer than most, and one in captivity lived for twenty years. Generally, as for humans, the female lives longer than the male. And yes, almost all spiders have some kind of poison. It is enough to paralyze or kill the prey, which are quite small (insects, for example) compared to people. The only spiders really dangerous to most humans are the black widow and the brown recluse, and even their bite is rarely fatal to man. Although the tarantula is poisonous, it seldom bites unless severely provoked. Like a number of spiders, the tarantula does not normally spin a web.

We might note that most spiders, even the notorious black widow, have fragile fangs that can break one's skin only with great difficulty. Many bites attributed to spiders are more likely bites from flies, ticks, mites, and other insects. There are few documented sources on spider bites. There is less than one chance in one hundred that a bite you receive is from a spider. Spiders have an undeserved reputation as being vicious; they are helpful in eating many insects that can destroy crops.

Q. *Why do snakes flick their tongues in and out?*

A. The snake's tongue is an important part of its sensory organs, a sort of combination of taste and smell. The tongue may pick up a particle from the air and transfer it to two depressions in the roof of the mouth where its meaning is further evaluated. Snakes usually have good eyesight as well but are deaf in our understanding of the word. They can sense vibrations passing through the ground. One reason you may not see many snakes in the woods is that they sense your foot vibrations and are long gone before you come within sight. Although snakes are often considered a symbol of evil or the Devil, as in the biblical account of Adam and Eve, they have their good points, a major one being that they eat rats and other rodents. This saves millions of dollars in agricultural products each year and cuts down on the spread of diseases carried by rodents. In their evolution, snakes at one time walked. The skeleton possesses little hip girdles that are vestigial.

Q. *I've heard that a chameleon can change its color even if it is blind. Is this true?*

A. This question implies that the chameleon can change its color at will to match its background, most likely as a protection against enemies. This is a widely believed myth. Indeed, the chameleon does change color to shades of brown and green, or yellow, sometimes with spots. This is brought about by the spreading out or concentration of pigments in cells. Color changes are brought about not by the will of the animal, but by automatic responses to changes in temperature or lighting in the environment. Sometimes color changes occur due to emotion, either fear or triumph. It can be compared with some humans who get red-faced when embarrassed. We can imagine that the chameleon occasionally escaped being eaten when a color change coincidentally matched its background of grass or rock, blind or not. The chameleon is mostly a tree-dweller, eats insects, and lays eggs. Each eye moves independently of the other.

Q. *What kind of animal is that little bear that Qantas Airways uses in its advertising?*

A. Although it is well known as the koala bear, it carries the fancy technical name of *Phascolarctos* and is a marsupial mammal more closely related to a kangaroo than a bear. As such, the mother of the species has a pouch to carry her young, but oddly enough the pouch is in back rather than in the front. The koala likes to live in trees rather than airplanes, and may eat up to three pounds of eucalyptus leaves per day. It is the only food they will eat. If there were no eucalyptus trees, the extinction of the koala would be certain and swift. Once hunted for its fur and threatened with extinction, these animals are now rigidly protected in Australia by law. They may live up to twenty years of age.

Qantas Airways, by the way, is one of the world's oldest airlines, having started operations down under about 1920. The acronym QANTAS stands for Queensland and Northern Territory Aerial Services, but of course nowadays it serves all continents.

Q. *Pandas seem to be an endangered species. Are they found only in China?*

A. It would be difficult to find a more endangered species than the panda because there are only 1,000 of them estimated to be left and they reproduce slowly. And yes, the south-central part of China seems to be their only stamping ground, but as we know, the panda has captured the hearts of children and adults everywhere. They live in mountainous forests where bamboo, the panda's principal food, flourishes. These areas have been encroached upon and destroyed by China's expanding population. Besides that, they have been hunted and killed for their attractive black-and-white luxuriant fur.

However, the Chinese government has imposed life sentences on anyone caught hunting the panda.

China has given more than one hundred pandas to zoos around the world, where strong efforts are being made to encourage reproduction. The pandas can grow to five feet tall and live to be twenty years old in captivity, perhaps fifteen years in the wild. Their usual weight is 220 pounds. An odd fact is that while they subsist chiefly on bamboo in the wild, eating up to 65 pounds of it a day, those in captivity do very well eating vegetables, cereal, and milk. There is a good chance the panda can be saved, as was the bison. Biologists still argue whether the panda belongs to the bear or the raccoon family. Some favor a separate taxonomic category altogether for the panda.

Q. *Scientists can train white mice to run through mazes. Have they been able to do that with lower creatures such as insects?*

A. There is growing evidence that insects can learn from their mistakes, contrary to the general belief that insects are instinctive and stupid. In experiments at the University of Illinois, praying mantises were fed a diet of milkweed bugs, which in turn had been fed either of two diets. One of these diets was a toxic substance and apparently tasted bad to the mantis, and so was rejected by them. Subsequently, they also refused the other group of milkweed bugs that had been raised on nontoxic sunflower seeds. The mantises also refused any insect disguised to look like a milkweed bug. There are many other examples showing that insects may have more smarts than we give them credit for. Yet as far as we know, they cannot master mazes with the same efficiency as mice. Those who claim insects will take over the world if humankind becomes extinct may not be too far off the mark.

Q. *Are there other animals that laugh? I know the hyena laughs. Maybe it is laughing at its own cowardice.*

A. The hyena laughs, or appears to, but not because it has just heard a funny joke. It is the natural call of the animal. It is very difficult for scientists who study this in the laboratory. It is easier to know more about fear and anger in animals because these reactions are readily induced with mild electric shocks. But how does a rat behave in a humorous situation, and how do we know if a rat is amused? As signs of an emotion, laughing and smiling are a part of what we call happiness. We have all seen dogs and cats at play and it is clear they are enjoying themselves, but are they laughing? Perhaps humans, by virtue of higher perception, intelligence, and reflectivity, have developed humor to a sharper edge than lower animals. In short, humans are probably unique in laughing over incongruous situations, which is the essence of humor.

Hyenas may be timid and retiring but they are not cowards. They can be formidable and dangerous if they are hungry or if their offspring are threatened. A major reason for the poor reputation of these animals lies in their nocturnal habits and their diet of the remains of kills of other creatures such as lions. They often follow the lion, eating what the lion leaves behind. Owing to this propensity to eat dead flesh in the darkness, these animals as scavengers have become associated with evil and the Devil. In addition, the laughing bark of the hyena can often sound human, and so stories have arisen of men changing into hyenas for evil purposes. However, hyenas have sometimes been known to lead lost persons safely out of the wilderness.

Q. *Do bears really sleep all winter? Why wouldn't they starve?*

A. While there are numerous insects, amphibians, and reptiles as well as mammals that hibernate, oddly enough the bear is not a true hibernator that remains dormant throughout the winter. There are many kinds of bears from the grizzly to the brown bear to the polar bear. They live in a very cold climate and gain fat during most of the year, and when winter arrives they sleep, awakening at irregular intervals to wander about, but eating very little. By spring, the bear's intestinal tract is in a state of partial collapse.

Bears are a diverse group, but they do like sweets such as honey, and are among the few wild animals that get tooth cavities. The polar bear is interesting in that many of them may live their entire lives without touching land, swimming and living on large chunks of ice. For protection and bearing of young, bears may make their den as a hole in the ice or a hole under a tree, or in a convenient cave. They are omnivores like humans, eating meat or other food, as it is available.

Q. *Is there any real difference between a hare and a rabbit?*

A. There is, even though they both belong to the same family that biologists call the *Leporidae*. At birth, the rabbit has no hair, its eyes are closed, and it is helpless. A hare is born with eyes open, a coat of fur, and it can hop around within minutes of being born. Also, a hare has longer ears. If you see a "rabbit" whose ears are longer than its head, it is a hare. *Jackrabbit* is another name for a hare; the name was coined because its ears resembled the long ears of a donkey (jackass/jackrabbit). As we all know, rabbits (and hares) multiply. This is due to the fact that up to eight in a litter of these rodents can be produced in thirty days from conception to birth. Thus a female rabbit, or doe, can produce several litters each year. Rabbits have been useful to humans as a source of food and as a source for felt hats and fur coats. However, they can destroy crops because of their voracious eating habits. In our analysis, it is the rabbit that brings more baskets of eggs to children at Easter than the hare.

Q. *How long can a camel go without water? Are they really stupid animals?*

A. We haven't conducted any IQ tests on camels lately, but they are truly remarkable animals. A camel can go for seventeen days without drinking any water. Carrying a load of 500 pounds, they can travel 75 miles over a three-day period, again without drinking any water. There is a secret to this. The camel carries a great deal of fat in its hump and has the ability to manufacture water out of this fat by oxidation. This is not to say the camel doesn't get thirsty. When it gets the chance to drink after a long drought, it can suck down 25 gallons of water. Camels can live forty years. Oddly enough, the camel originated in North America, where it no longer exists. We might add that camels have very pretty eyelashes, but don't look too closely or the camel will spit at you.

Q. *Are there any authentic cases of packs of wolves killing and eating humans?*

A. We know of none. Human flesh is not a normal part of the wolf's menu although they have voracious appetites. The notion of a pack of twenty or more wolves chasing sleighs or surrounding the trapper's cabin in a snow-swept forest is pure fiction. Timber wolves are the common species and they keep to themselves in groups of five or six. True, they have a fondness for sheep and other domestic animals but their main diet consists of rodents and rabbits. On occasion, they are capable of chasing down larger prey such as deer, elk, or horse. The idea that wolves eat humans may have derived from the sight of wolves sniffing around corpses on battlefields in earlier times when wolves were abundant. If anything, the wolf has a beef against us. Humans have exterminated the wolf in the British Isles and western Europe, and nearly so in North America. In places where the wolf has attacked sheep or other domesticated herds, ranchers have hunted and killed them via airplane. While the wolf is most often viewed as a symbol of ferocity, there are many stories wherein the wolf is friendly.

Q. *A deer hunter friend of mine says that deer have no gall bladders. How can they live without them?*

A. Not only deer, but rats, horses, and even pigeons have no gall bladders. We might say it is useful but not essential for survival of the organism. Basically the gall bladder is a storage place for bile. Bile is a greenish or yellowish alkaline fluid secreted by the liver and aids in the digestion, especially of fats. In humans, if no food is in the small intestine, the body will save up the bile in the gall bladder until something is eaten. The gall bladder can store nearly two fluid ounces of concentrated bile. In some people this can lead to trouble if the bile crystallizes to form gallstones. Fortunately, the gall bladder can be removed surgically and the person will function normally. In deer and some other animals the bile moves directly from the liver to the digestive tract. While deer may escape the problem of gallstones, they can get arthritis just like we do.

Q. *Is it true that the giraffe is the only animal that has no voice?*

A. No. The giraffe makes a low moan as its call, but it is seldom heard, and so the popular idea persisted that the giraffe is totally silent. Nevertheless, it is a remarkable, graceful animal, being the tallest mammal alive. It subsists almost exclusively on acacia leaves, swallowing them with the help of an 18-inch tongue. Although the giraffe appears ungainly, it can move along at 30 mph and defend itself well with kicks from its strong legs. Nonetheless, the lion remains its chief enemy, next to humans. The notion that the long neck of the giraffe is the result of stretching for treetop vegetation is without any scientific foundation. During the nineteenth century, some scientists held that acquired characteristics, such as a lengthened neck, could be passed on to offspring. In other words, they thought traits not used would disappear. This was called the *law of use and disuse*. Scientists have discarded the concept due to lack of evidence.

Q. *Why are foxes regarded as especially clever?*

A. Members of the dog family, which includes foxes, are intelligent. In addition, foxes have above-average hearing, sight, and smell. These advantages can make any animal look good, as when being pursued by hunting dogs. There are a variety of foxes, but the most common are the red fox and the gray fox. The red fox seems to be smarter, but the gray fox knows how to climb trees. They are generally shy, retiring animals that live mainly on rodents and rabbits but will take a farmer's chicken if given the chance. Actually, humans are the fox's chief enemy since the fur from such varieties as the silver fox is considered by some to be desirable. Yet the fox helps humans by eating destructive rats and mice. Foxes can be domesticated.

Q. *Why do cats purr?*

A. In our experience, cats that are purring seem to be pleased with themselves, but they are also known to purr when in extreme pain. Many authorities agree that purring, as a vibratory signal, is a homing device for kittens to come to the mother for milk. Not all members of the cat family purr. Lions and other big cats roar instead. There is a widespread notion, while we are at it, that cats hate water and cannot swim. Many cats indeed like water and do swim quite well. What they don't like is being suddenly doused or immersed in cold water, an attitude we can all relate to. Additionally, one member of the cat family, the cheetah, is the fastest mammal alive. It can attain short bursts of speed up to 70 mph.

Q. *If white elephants exist, why are they considered bad luck?*

A. Yes, they exist, but are rare. They are albino elephants found mostly in Thailand and Burma. They are more of a grayish-white with pinkish eyes rather than a more startling white. The great showman of the nineteenth century, P. T. Barnum, in his efforts to bring to the public the most unusual of creatures for his famous circus, obtained at great cost a white elephant for exhibit. Such an elephant, albino or not, is expensive to maintain, for it eats up to 500 pounds of food per day and drinks 50 gallons of water. Unfortunately for Barnum, the white elephant was disappointing as an attraction for the public. Barnum not only lost money, but also was unable to find anyone who would buy the animal. Hence, the expression came into our vocabulary of having a "white elephant," something expensive to keep and hard to get rid of. We have not found out how Barnum finally got rid of the beast.

Q. *Can a bat really get tangled in my hair?*

A. This is a myth. With its echo-sounding system, a bat can easily avoid not only your hair, but the rest of you as well. Perhaps the basis for this myth is that when a bat hibernates, it becomes sluggish and not in full control of its flying ability. Under such circumstances, a disturbed bat may on occasion strike someone in the head. In our Western culture, we associate the bat with graveyards, darkness, and evil things. Demons have bat wings, while angels don't. Yet, oddly enough, the Chinese regard bats as symbols of happiness and longevity. This is logical because for a small mammal, bats have long lives—some have lived up to twenty-one years. Another thing about bats that isn't generally known is that their homing instinct rivals that of the homing pigeon. The little brown bat has been banded and shown to fly sixty miles to its home in a single night. And it can do it blindfolded! One other nice thing we like about bats is that they can eat as many as one thousand mosquitoes in a single night.

Q. *How do beekeepers handle hives and manage not to be stung?*

A. Sometimes they do get stung, but some experienced beekeepers prefer to work without protective gloves. They wear a veil over their heads and shoulders and generally employ a smoking device with a bellows to quiet the bees. Beekeepers have also learned that slow, deliberate movements will not excite them. Also, the bees chosen to inhabit the man-made hives are those that produce the most honey and are the most docile. A single hive may produce 30 to 40 pounds of honey in a season. Beekeepers have more to worry about than getting stung. They have to guard against predators such as skunks, toads, and mice that invade the hives not only for the honey but to eat the bees as well.

    The nectar gathered by bees from flowers is converted to honey in the honey sacs of the bees. In the process, sucrose is converted to fructose and dextrose, so honey is essentially sugar. On the other hand, royal jelly is secreted directly from the salivary glands of the worker bees and fed to certain female bees, which transforms them into queens (they are not born as queens). Royal jelly is nutritious, containing carbohydrate, protein, and even vitamins. Beeswax, a by-product, has many uses besides honeycombs. It is used in floor wax, wax paper, and candles. Some religious rites call for the use of beeswax.

Q. *Why do I have so many fruit flies? They seem to come out of nowhere.*

A. What we tend to call a fruit fly isn't a true fruit fly at all but rather a vinegar or pomace fly with the Latin name *Drosophila*. It does, however, lay a large number of eggs on fermenting or ripe fruit, and the parents can produce thousands of offspring in a short time—short because these little flies have a life cycle of only two weeks. *Drosophila* and the true fruit flies, especially the Mediterranean fruit fly, have caused considerable damage to citrus crops. Nonetheless, *Drosophila* is very useful to scientists because of its short life cycle and large chromosomes, which are easy to see and study. From the study of these flies, we have gained considerable knowledge of genetics and the workings of evolution.

Q. *Do any birds have teeth?*

A. There are no birds living today that have teeth, though some have toothlike projections or notches along the mandible, but these are not true teeth. Actually, the first birds did have teeth and were closely related to the dinosaurs. The first bird was about the size of a crow and had a full set of teeth. Fossilized specimens were found in southern Germany under such excellent conditions of preservation that the feathers could be seen. Otherwise, the remains would have been classified as a reptile. The name given to this creature was *Archaeopteryx*. From that beginning about 150 million years ago, birds flourished and grew larger, and most of them had teeth, including a few bigger than the ostrich. One species laid an egg fully 13 inches long. The success of birds is underscored by the fact that there are twenty-five birds for every person on this planet, or more than 100 billion birds. Despite this, there are birds that are threatened with extinction, such as the whooping crane and the California condor. The passenger pigeon used to darken the skies but is now extinct.

Q. *What is the fastest bird alive in terms of speed in flight?*

A. The debate continues as we investigate further into this question. Some say the peregrine falcon is the fastest, timed at 180 mph in a dive. Others argue that the swift is the fastest. Two spine-tailed swifts have been clocked at 172 and 218 mph, but under unknown circumstances. Another candidate is the frigatebird, reaching speeds of 160 mph. However, many reports of flight speed seem to be exaggerated unless they are moving with and propelled by a strong high wind. Birds also fly faster when in danger or frightened, or if they are hungry.

Frigatebirds derive their name from the habit of plundering fish from other birds in midair, although they are perfectly capable of catching their own food. The term *frigate* refers to the swift sailing warships of the seventeenth and eighteenth centuries. Polynesians have used frigatebirds as homing pigeons to carry messages. With its 8-foot wingspan, it is fast enough, but we won't know for sure until we enter prospective candidates into the Olympics.

Q. *I sometimes hear the expression "dead as a dodo." Isn't this some kind of bird?*

A. It is an extinct bird distantly related to the pigeon. It once inhabited the islands of Mauritius and Reunion in the Indian Ocean, east of Africa. It is another of those creatures whose contact with man led to its extinction. When settlers arrived on those islands, they brought pigs, which apparently feasted on the eggs of the dodo. This led to their permanent demise sometime during the seventeenth century. The dodo was about the size of a turkey, could not fly, and was somewhat clumsy. Although living specimens were sent to Europe, there is today no complete bird on display in any museum that we know of in Europe. However, there is one specimen at the American Museum of Natural History in New York City.

Q. *Do ostriches really hide their heads in the sand when frightened?*

A. Not unless it is an extremely cowardly ostrich and even then we doubt it. More probable are faulty observations and unwarranted conclusions about the behavior of ostriches. The ostrich must arch its long slender neck in order to reach food on the ground or in bushes, which led to the notion that the bird was trying to stick its head in the sand to hide. In addition, the ostrich often sits on the ground with its legs folded and its neck and head stretched out along the ground observing the surroundings with its keen eyesight. To a distant watcher, only the body was visible and the head and neck appeared to be buried. The ostrich is actually a noteworthy bird, the only one with two toes, and is known for laying a whopping 3-pound egg. It is the largest living bird, and if frightened can leave any unpleasant scene at a graceful 40 mph.

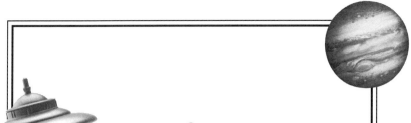

9

THE GREEN EARTH
Plants

Q. *What are the major differences between plants and animals?*

A. Plants derive energy by photosynthesis and the presence of chlorophyll. Also, plants are rooted and have a cellulose support for the plant's structure. Animals, on the other hand, derive their energy by ingestion and assimilation of organic material, i.e., plants and other animals. They are usually unattached, mobile, and are usually supported by a bony external or internal skeleton.

There are common grounds as well, however. Many species of plants, such as the Venus's-flytrap, are carnivorous. Some are mobile, as with phytoplankton in the oceans. Plants may often be male and female and not asexual. Many animals are attached or *rooted*—clams, oysters, and corals are examples. In addition, there are bacteria, viruses, and other forms that defy classification as either plants or animals. Yet most of the time, animals and plants can be differentiated if you remember that most plants are green.

Q. *Are the giant redwood trees found only in California?*

A. Although the redwoods extend somewhat into southern Oregon, California has a virtual monopoly on this magnificent tree. The closest relative we could identify is the Japanese cedar in the Far East. This tree grows to a height of 150 feet with a circumference of 25 feet.

There are two species of the *Sequoia*, known commonly as the *redwood* and *big tree*. Both can be seen in either Redwood National Park or Sequoia National Park. At Sequoia, the most impressive tree is the General Sherman Tree. It is 272 feet high with a diameter of 32 feet and a circumference of 101 feet. The tree contains more than 6,000 tons of wood. Other trees exceed 300 feet in height but are more slender. The General Sherman Tree is about 4,000 years old, the oldest living thing next to the bristlecone pine. These giant trees used to be widespread across the Northern Hemisphere, but that was 150 million years ago. Strangely, the wood is useless as timber. It is brittle and shatters into splintery irregular pieces when struck. Perhaps this is why it still survives.

We might add that the name *Sequoia* was given to the tree in honor of the Cherokee Indian Sequoyah, who developed the written language of the Cherokee.

Q. *Can plants grow from seeds hundreds of years old?*

A. Yes, although the longevity of seeds varies widely from one plant type to another. Some seeds may last only days or weeks before losing the power to germinate while others are good for years. In fact, seeds of the Oriental lotus found in a peat bog in Manchuria, dated at about 1,000 years, produced flowers. It is of importance that seeds be distributed some distance before taking root to spread and avoid overcrowding. The wind serves as a useful distributing agent. Grass seed has been found at heights of 3,000 feet, and of course everyone is familiar with the little parachutes of the dandelion. Seeds also spread via water, such as floating coconuts, and other seeds are carried to new areas by animals, including birds.

Q. *What has been the chief cause of rain forest destruction?*

A. The marked shrinkage of rain forest has been brought about chiefly by human activity. The Amazon Basin in South America is notably affected because of removal of timber, especially mahogany and cedar. This is accompanied by clearing for human habitation and grazing land for cattle and other domesticated animals. The need for living space is acute in Brazil, which is 40 percent rain forest, due to an expanding population. The result is that the rain forest, as a great ecosystem girdling the equatorial belt of Earth, is the most threatened of ecosystems in terms of size reduction and extinctions of the prolifically diverse plants and animals that harbor therein.

Rain forests survive best in hot, wet climates where rainfall, the key factor, exceeds 70 inches annually. In a way, the rain forests are a leftover from the time of the dinosaurs when the climate was hot and steamy, because these areas include some of the most primitive vegetation on Earth. However, later cooling caused the rain forests to retreat to the equatorial regions. Of note is that the *angiosperms*—that group which includes the flowering plants, fruits, and nuts—originated in the rain forests. The rise in importance of mammals coincided with the spread of the angiosperms about 60 million years ago.

Q. *Where does cork, such as that used as bottle stoppers, come from?*

A. Cork is the bark from a type of oak tree found in the Mediterranean area, especially in Spain, Portugal, and North Africa. Cork is filled with air bubbles that provide insulation for the oak tree against the changing weather. These trees don't produce much cork until they are more than fifty years old. When the cork is cut, say, to make stoppers for bottles, the air bubbles are also cut, creating little suction cups that help them to hold tight against the bottle. Because cork has only one-fifth the density of water, it is just as useful in life preservers as in bottle stoppers. It also has excellent insulation and sound-proofing properties.

Q. *The three wise men brought gold, incense, and myrrh. What is myrrh? Is it valuable?*

A. It is a gum resin exuded from the myrrh tree, which is a short, stubby tree 4 to 20 feet high found in Somalia and some Middle East countries. After the myrrh oozes out, it hardens into small irregular lumps. Many centuries ago, Europeans sought after it for use as a perfume. However, it has a bitter taste; in fact, the Arabian word *murr* means "bitter." It was also used in medicine as an antiseptic and for embalming. Its value today is slight. One novel use is in Christmas tree ornaments where a few small lumps of myrrh are enclosed inside clear plastic balls.

Q. *I like cashews and was wondering if they grow the same way peanuts do.*

A. Unlike the peanut, the cashew grows on trees that may reach heights of 40 feet or so. The tree, which originated in Central and South America, grows best in tropical or subtropical areas. During the fifteenth century, missionaries brought it from the New World to East Africa and India. The cashews you eat are probably from India, and the cost of importation makes them relatively expensive.

If you can imagine a fruit the size of an apricot with a cashew protruding at one end, this is what the cashew looks like as it comes off the tree. The fruit, or *apple*, can be used to make jams and jellies. The cashew has a double-walled shell, which contains irritating oil that can cause skin eruptions. This is because, unknown to most cashew lovers, the cashew is a member of the poison ivy family. Special roasting ovens have gotten around this problem. Yet the noxious oil has commercial use as an insecticide and as a lubricant. Some plastics are also made with it. About one-half million tons of cashews are produced annually. By the way, the peanut is not a true nut. Nor is the coconut or the Brazil nut.

Q. *Where are lemon trees found?*

A. They are found throughout the Mediterranean area in Italy and North Africa and also in the southwestern United States. These are mostly cultivated trees 15 to 25 feet high that produce rather incredible numbers of lemons. Some trees may produce up to seven thousand lemons in a single year. The average is about fifteen hundred. The origin of the lemon tree is not known for certain. The early Crusaders found them in Palestine and brought them back to Europe. They are rare in China and India. Italy is today a world leader in lemon production, but oddly enough, the ancient Romans didn't have any lemons. In some cultures, the lemon is thought to have magical properties. By driving several iron spikes through a lemon, you can ward off the evil eye, if you need to do this.

The taste and smell of a lemon comes from the organic compounds aldehyde and ester, found in oil glands of the peel. After the juice is extracted, it is concentrated by evaporation in ratios of 3:1 to 6:1. Lemons are rich in vitamin C, which is preserved by passing the freshly squeezed juice into a tank maintained as a vacuum; otherwise, the vitamin C would be destroyed by oxidation. One of the main uses of lemon is in making lemonade, and as might be expected, the hotter the summer, the more lemons are sold. We might speculate that the lemon's reputation as a "refresher" accounts for its popularity in such products as soap, perfume, and furniture polish.

Q. *Where does chocolate come from?*

A. Chocolate is the most popular candy and bakery product in the United States, and we consume more of it than other country. It comes from the cacao (or cocoa) tree, which grows in tropical areas and can attain a height of 40 feet. These trees produce orange-yellow to reddish pods resembling cucumbers which when ripe can be harvested and opened up. Inside, there are twenty-five to fifty almond-shaped seeds or beans which when roasted and ground up are the raw chocolate. This concentrate can then be treated with vanilla, milk, sugar, and other additives to produce the varieties of chocolate we know.

The Aztecs and Mayans used chocolate for centuries, including the use of the bean as money. When Columbus and Cortez came to America, they brought the cacao pods back to Spain where it remained a jealous secret and precious beverage for more than one hundred years. However, the French and English obtained it and added sugar and milk to perfect the flavor. It was so expensive that only the rich could afford it. It wasn't until the middle of the nineteenth century that the use of chocolate spread worldwide. In an average chocolate bar there is as much sugar as chocolate, and if you eat one pound of chocolate, you have consumed about 2,200 calories. Yet, chocolate contains about 18 percent protein.

Q. *How long can an orange tree produce oranges?*

A. An average orange tree will produce fruit for fifty years, but eighty years of productivity is not uncommon, and a few trees are known to be still producing fruit after more than a century. An orange tree may attain a height of 20 feet, but some trees are as much as 30 feet high. Trees grow well in a variety of soils in generally subtropical settings. While orange growers fear frost, a light frost may be beneficial in terms of taste and firmness. The orange has been known for several thousand years, originating in Southeast Asia and spreading from there to India, Africa, and the eastern Mediterranean area. Yet today, the United States is the leading producer of oranges. Tangerines are a variety of orange, and a tangelo is a cross between a tangerine and a grapefruit.

World production exceeds 36 million tons. This translates to about 238 billion oranges each year, sufficient to supply forty-eight oranges to each person in the world. However, 40 percent of the production goes to make frozen concentrate. An old English superstition suggests the orange as a way to gain the attraction of a young lady. The suitor pricks the orange with a needle in several places and sleeps with the orange under his armpit, and then the lady must eat the orange. A bit unhygienic, but maybe it works.

Q. *How extensive was Luther Burbank's contribution to plant science?*

A. It could be said rightly that he was the foundation on which the breeding and development of new and beneficial strains of fruits, grains, vegetables, and even flowers was built. Burbank was born in 1849 on a farm in Massachusetts, and never attended college. Early on, he developed an improved potato, and then moved to

California where he spent most of his career. Burbank developed more than eight hundred new strains before his death in 1926.

We would interject here that many of the things we eat such as tomatoes, corn, and plums in their original condition were almost inedible by our standards today. Tomatoes weren't even red. Only through patient selection, grafting, and hybridization were new, tastier, and more nutritious strains produced. Many of Burbank's strains of prunes and plums are of commercial value even today. In 1923, Rudolph Boysen created the boysenberry by multiple crossings of blackberries, a raspberry, and a loganberry. Such efforts have led to the Green Revolution and the production of superior crops that have saved millions from starvation.

Yet when Burbank published his seed catalogue at the turn of the century, pointing out the advantages of his hybrid seeds, he was vigorously attacked from the pulpits of churches for blasphemy because he had infringed upon God's right to be the only creator. Burbank, California is not named in honor of Luther Burbank, but rather David Burbank, a Los Angeles dentist.

Q. *What is the most nutritional fruit? And what is the difference between a fruit and a vegetable?*

A. The avocado is probably the most nutritious because it contains eleven vitamins including A, C, and E; up to 25 percent unsaturated fat; and about 740 calories per pound. The avocado is native to the Americas and was long cultivated in pre-Columbian times. Although some people consider it too bland, it can be pepped up as guacamole by mashing the fruit and adding grated onion and lemon juice with a dash of tomato pulp and parsley—good for stuffing celery or on tortilla chips. A little salt may help. In contrast, the cucumber, tasty sprinkled with vinegar, is the least nutritious, with only 73 calories per pound.

Even an unabridged dictionary lists the tomato as a fruit and a vegetable. In general, a fruit is anything edible developed from a seed or developed from a flower. Vegetables include anything from herbaceous plants that you can eat, such as leaves and roots. Thus, potatoes and beets and other tubers would have to be vegetables. At times, the distinction is not very sharp.

Q. *Is the tomato a fruit or a vegetable?*

A. Take your pick. A biologist would say it is a fruit, but in a Supreme Court decision (this sounds silly) in 1893, it was classified as a vegetable because it was typically served and eaten with other vegetables. So much for science. The tomato has an interesting history. It seems to have been a wild yellow species found in Bolivia and Peru, and then cultivated in Mexico and shipped to Europe after Columbus landed. The Italians called it the Golden Apple because of its yellow color, but soon scarlet varieties emerged. In the United States, it seems to have been first grown by Thomas Jefferson in 1781, but a lot of people refused to eat it until as late as 1900 because it was believed by many to be poisonous. For many Europeans long ago, it was the *love apple* because it was thought to make the person more romantic. Tomatoes are of course neither poisonous nor romance enhancers, but they are excellent sources of vitamins A and C.

Q. *The flesh of a watermelon I cut into wasn't red but a bright orange. Was this a freak watermelon?*

A. No, it was not, and we'll bet you found it delicious, assuming you ate it. People are so used to the typical red-fleshed watermelon in the supermarket that they forget, or may not know, that there are several varieties ranging from shades of white to red to yellow. Actually, the watermelon is a member of the gourd family and is thus related to pumpkins, cucumbers, and squashes. Some watermelons may weigh 50 pounds. They have been grown for at least 4,000 years, being depicted in murals from ancient Egypt.

Q. *Who discovered how to make wine from grapes?*

A. We do not know who, but there is a good chance it started in the Mediterranean or the Middle East, perhaps among the Greeks or Egyptians, or the early inhabitants of Italy. Its discovery may well have been accidental. We can visualize some grape juice "spoiling" and then being consumed. Thereafter, its preparation would have been deliberate. Early wines must have tasted awful by our standards today. Wine was aged in goatskins stoppered with greasy rags. Air could still get in, which is undesirable. It wasn't until the eighteenth century, when bottles and corks came into wide use, that wine could be aged properly. Good or bad, wines were drunk thousands of years ago, as mentioned in the Old Testament, and we know the ancient Egyptians drank beer more than 5,000 years ago during the pyramid-building epoch. Wine probably came before beer, since wine can form without human intervention, but beer has to be made. The development of winemaking as an art was fostered by early monks and priests because wine was, and still is, used in Communion services.

Q. *Why are some mushrooms called* **toadstools**?

A. The term *toadstools* refers to mushrooms that are inedible or poisonous. When a toad is alarmed or attacked, the warts on its back secrete *bufonenin*, a poisonous substance if taken internally. Curiously, this is the same chemical produced by the warts of the poisonous mushroom *muscaria*. Perhaps that is the connection. Mushrooms have been eaten for thousands of years and were considered a delicacy by the Greeks and Romans. Ancient priests were so possessive of mushrooms that they forbade ordinary people to eat them. Although tasty, mushrooms have little food value and may consist of 90 percent water. There are many varieties of edible mushrooms but also a number of poisonous ones that can actually be fatal. One who collects and eats mushrooms indiscriminately is asking for trouble. Some mushrooms are hallucinatory.

Q. *When did man begin cultivating and eating the potato?*

A. The best estimate we have is about 1,800 years ago in the Peruvian Andes of South America, but it may have been earlier. When the Spaniards invaded Mexico and Peru in the sixteenth century, they brought back the potato to Europe. At first, Europeans were suspicious of the potato and even believed it caused syphilis and other diseases. However, it soon spread to Germany and Ireland and became established as a staple. The Irish became so dependent upon the potato that serious famine resulted in 1845–46, when the crop failed due to blight.

There are varieties of potato that have yellow and purple flesh rather than the usual white. Whatever the color, potatoes are a good source of vitamin C and contain protein and thiamin. Also, they have about the same low number of calories as an apple.

Q. *Are there other plants that eat insects besides the Venus's-flytrap?*

A. Yes, the Venus's-flytrap is only one of about four hundred species of carnivorous plants. They have a wide range and variety. They are found both in tropical and temperate climates from Australia to South America and other places. Many resemble ordinary flowering plants and few look like the Venus's-flytrap. For most of these unusual plants, the ingestion of insects is a dietary supplement because they still rely on photosynthesis to make food.

It seems that carnivorous plants evolved in wet or otherwise swampy areas where nitrogen was in short supply. When digested by enzymes of the plant, insect bodies yield a great deal of nitrogen. Some plants have hairlike tentacles that secrete a sticky substance. When a fly or other insect lands, it is entrapped and engulfed by the sticky tentacles. So it was nature that invented flypaper.

10

NUTRITION AND HEALTH

Q. *We are urged to eat a "balanced diet," but what does, say, an apple do for you that a steak doesn't?*

A. What we eat—the fuel that keeps us alive—can be divided into different food groups. The principle groups are proteins, fats, and carbohydrates, in addition to water, vitamins, and minerals. An apple is a carbohydrate and steak is protein with some fat. The balanced diet assures that all the elements needed for good nutrition will be available for utilization by the body. It is from the processing of nutrients that good health flows.

Protein is of critical importance to the body. It consists of chainlike sequences of amino acids whose role is to build tissue and provide energy for the organism.

Ingested fat can be used immediately as a source of energy or serve as stored energy until needed. Carbohydrates provide quick energy, as sugars, and are the true brain food (rather than fish) because the brain must have glucose to function. Carbohydrates can be converted to fat; and proteins, if necessary, can take over many of the functions of other food groups.

We think of ourselves as solid bodies but we are really vessels of liquid. This is good because chemical processes take place chiefly in liquids, and our bodies are busy places of constant chemical reactions necessary for life. This is why we must have water. As for vitamins and minerals, these act as regulators and facilitators of chemical functions within the body but do not provide any energy on their own. They safeguard us from many diseases and abnormalities of growth and development. Without vitamins and minerals, we could not survive. It can be seen that the nutritional process in our bodies is incredibly complex, but if we eat the right foods, nature usually gets the job done.

Q. *We are constantly bombarded with food products proclaiming they are fat-free or have less fat. One would think fats are poisonous.*

A. Perhaps a major motive for shoppers looking for foods with less fat is the desire to look thin rather than plump, and to derive better health in the process. Far from being poisonous, fat is a major and essential food group. Its importance can be seen when there is a fat deficiency. In that case, protein and carbohydrates will reorganize and create fat to make up the loss. Fat provides immediate usable energy; some is stored for long-term use. Fat is also a good insulator from cold as we see in the layers of fat possessed by seals, whales, and bears. Without fat, the body would be unable to absorb the fat-soluble vitamins. Fat makes many foods tastier, such as pastry, candies, and steak.

On the debit side, too much fat can cause obesity, high blood pressure, clogging of the blood vessels, heart problems, and stroke. This is particularly true of saturated fats. These fats, such as shortening, are solid at room temperature. Unsaturated fats, such as olive oil and peanut oil, are liquid at room temperature. Nutritionists recommend a diet with no more than 25 to 30 percent fat. It might be noted in passing that cholesterol is an organic compound but it is not fat. It is present in fat, however, and excessive amounts contribute to hardening of the arteries. Cholesterol is manufactured by the liver and is a key player in cell membrane construction, hormone function, and bile production.

Q. *There are dozens of diets saying you can lose weight. Do any of them work?*

A. The many varieties of diet books is a response to the popular demand to lose weight painlessly. In some cases, people are trying to gain weight. For many suggested diets, success is achieved if the diet is rigidly adhered to. Most diets fail due to human nature and lifestyle. Often, after the first flush of enthusiasm, the dieter slowly reverts back to the old eating habits. A number of diets focus on a single solution such as eat mostly meat or no meat at all; or eat anything but fat. Such diets are not only unworkable, but are also deleterious to good nutrition. Some diets rely on a catchy and easily remembered regime such as "eat nothing white." Such a diet would rule out bread, potatoes, milk, some salad dressings and cheeses, and most fish. A person might lose weight but starve to death in the process.

It is known that weight loss from crash diets is quickly regained. More permanent weight loss is achieved by small losses over a period of several months. A better program is to eat a normal balanced diet and in smaller portions. Make small substitutions of less fatty foods so that usual eating habits are not drastically altered. Appropriate exercise should accompany any diet.

Q. *Why is a vegetarian diet more healthful?*

A. It is not necessarily more healthful. Some vegetarians may require dietary supplements to overcome protein deficiency resulting from a too rigid vegetarian diet. As a modification, some vegetarians called *lactovegetarians*, consume animal products such as milk and cheese. The tooth structure in the human species is that of an omnivore, one who consumes or is designed by nature to consume both flesh and plant material. A vegetarian diet is thus unnatural.

Vegetarianism had its origins about 2,500 years ago in India and the Middle East. A chief motivation was the philosophy that fellow animals should not be killed. However, some individuals ate meat but only if they did not themselves kill the animal. Others were mainly vegetarian because meat was either scarce or too expensive. For example, during times of game scarcity in old Britain, they ate sharp cheese on bread and called it, with a sad smile, *Welsh Rabbit*. It was a case of forced vegetarianism, but we still eat it today under the name *Welsh Rarebit*. Vegetarians were also among those opposed to bloody sacrifices on ethical grounds. Another angle was reincarnation of human form to lower animals. No one would want to kill and eat a reembodied friend or relative. Subscribing to vegetarianism is not a guarantee of kindness toward other living creatures. Adolf Hitler was a strict vegetarian and suffered considerable flatulence as a result, which mightily embarrassed him.

Q. *I have a friend who says your own body will tell you if something is wrong with you and usually will take care of itself. How much truth is there to this?*

A. It is not a wise philosophy. It is true that the body gives warning signals for many ailments. Dizziness, edema, trembling, blurred vision, and numbness are all signals of disorder, but they are common to a number of diseases, each requiring different treatment. The body is unable to diagnose which disease is the culprit, and usually neither can the patient. The better signals offered by your body include pain, as you will more likely seek help than if there is no pain. Pain is often a protection, causing the body to recoil from a source of danger or damage such as fire or poison.

At the same time, the body may be deteriorating from a disease but the person is asymptomatic, until perhaps it is too late. Yet often the body heals itself, from small cuts and bruises to more serious problems. Autopsies reveal that, over a long life, some persons were beset by multiple disorders which the body managed to heal or control. However, the body is still unable to remove an inflamed appendix, properly set broken bones, or administer correct dosages of drugs not naturally occurring in the body.

Q. *I hear that malnutrition is the leading cause of death in Third World countries. Is this simply due to lack of food?*

A. Not entirely. Certainly, without any food, the body will start to consume itself, starting with stored fat reserves and then needed tissue, until survival functions cease and death follows. But it has been said, in truth, that one can seem to eat well, surrounded by plenty, and still suffer from malnutrition. One may eat several pounds of watermelon, cucumbers, and mushrooms per day and have malnutrition because these foods are mainly water. In Third World countries, or developing nations, the leading cause of malnutrition seems to be protein-deficient diets such as in Africa where infants and children suffer widespread *kwashiorkor*—severe malnutrition caused by a starchy diet of cassava after they have been weaned. This results in weakness, the potbelly of edema, and lack of resistance to lethal diseases. The solution seems simple because a supplement of dried skim milk, which contains protein, will cure this disease. It is nonetheless difficult to distribute in countries having political and military conflict, and with the remoteness of many villages, few roads, a dearth of health workers, and lack of education in nutritional matters among the populace.

Another cause of malnutrition, even in progressive countries, is a specific lack of one item in the diet. Without vitamin C, a person may get scurvy, which will cause teeth to fall out; without vitamin D, the bone disorder rickets results. This is why a broader diet encompassing all the bodily requirements is necessary.

Q. *I see that many people who lived centuries ago lived to ripe old ages. They probably didn't take vitamins and mineral supplements. Could our modern emphasis on nutrition be overblown?*

A. Not really. We can be impressed with dramatic statistics showing that the average life span in the days of the Roman Empire was twenty-five years while today it is about seventy-two years. However, one should be mindful that in those earlier times childhood diseases were rampant with high mortality and the younger ages were averaged into the age statistics in many cases. If a person survived the childhood diseases, the odds were good for living almost as long as we do today. But people today do have an edge in living longer because of improved sanitation and nutrition, as well as developments in medicine.

One natural advantage earlier peoples had, and that we share, is the penchant for the body to crave something it requires. For example, along the equatorial belt where rainfall is high, naturally occurring salt deposits are rare. When presented with salt, the people who dwell there will eat it like candy. Perhaps also, people in the Middle Ages or earlier instinctively consumed foods that contained needed nutrients. Another factor is that we are coming to recognize that a person's genetic makeup plays a significant role in longevity.

Q. *How did people keep food from spoiling before the days of modern refrigeration?*

A. The earliest methods we know of to preserve food include dehydration by drying it in the sun, salting it to prolong edibility, storing it in cool places such as caves, pickling, or making pemmican. Pemmican was made by pounding strips of meat vigorously (sometimes into a powder), adding lard and fruit as binders, and drying it to make strips or cakes. It was a highly concentrated food form valued particularly by hunters and explorers who wished to travel lightly. Another way to preserve food was by fermentation. This came about in Mesopotamia and Egypt by the 3rd millennium B.C.E. Colonial settlers dug root cellars, and during the winter harvested blocks of ice for the root cellars and packed straw around the ice for insulation. Then, during the summer months, food that spoiled easily was stored in these "ice boxes" underground. Another tactic was to divert cold meltwater streams fed by higher-altitude snow-covered areas so that the streams passed close to the house and food could be conveniently stored in the nearby water. In some cases, the stream, if small enough, was actually diverted under the house, providing the residents with basement refrigeration.

Of all the appliances in a household, it can be argued that the "fridge" is the one we are least able to do without. The evolution of the modern refrigerator began in the mid-nineteenth century by investigators in several countries. The fundamental principle was known: an expanding gas removes heat from the surrounding air.

This led to the vapor-compressor refrigerator in which a piston compresses a gas into a liquid (nowadays, freon) which then expands as it makes its way through a series of pipes back to the compressor, and cooling the air in the refrigerator during its journey.

Q. *How can people in some places around the world eat a diet of insects? The whole idea is repulsive.*

A. Even in this country we have a few "gourmet" types of people who like chocolate-covered ants and crispy grasshoppers. There are two points to keep in mind: (1) insects are high in protein actually and therefore do constitute a "food," and (2) food preferences are individual and cultural. If you had been raised in northern Japan, there is a white wormlike grub found under logs that you would have thought a delicacy perfect for barbecuing. Yet Western peoples are repelled by this. Some societies gag at the thought that we in the United States like big juicy steaks. In some South American countries, the thought of drinking a glass of milk would be as repugnant as the thought of drinking a glass of blood! On an individual basis, think of something a friend eats that absolutely repels you, such as ketchup on scrambled eggs. They would in turn become sick at the thought of you liking chocolate syrup on raw oysters. We all have different food preferences.

Keep one thing in mind: if you are starving to death, a few insects might save your life. Some hunters in a remote part of northern Canada were found starved to death at their encampment. The men were in a circle, huddled in blankets around a large gas lantern. The ground surrounding the lantern was littered with the bodies of large moths that had been attracted to the light. If they had eaten those moths, they may well have survived.

Q. *Is there any scientific reason to suppose that your state of health is reflected in your eyes?*

A. It would be nice to think that the human eye could work like the engine analyzer in a garage, pinpointing the trouble at the distributor or carburetor, and in this case, whatever diseased organ is causing trouble. This notion that the iris, or colored part of the eye, records the state of the body is called *iridology*.

Unfortunately, such a metering system doesn't exist, but as with other pseudoscience fads, it draws a number of practitioners and other believers. Its success is based on popular wishful thinking. In essence, who would choose exploratory surgery, if instead, simply reading the signs in your eye could discover the underlying disease? Iridology has been around as a so-called clinical technique for over one hundred years yet there has never been any well-documented evidence to support the claim that the eye reveals all of the disorders of the inner body. However, there are certain exceptions. A yellowing in the eye indicates jaundice. Blurred vision may indicate diabetes. Drug influence can be seen. In the case of a stroke, one pupil may be more dilated than the other. But these are symptoms that should be used in conjunction with other procedures to determine cause and treatment.

Q. *Why do we get pimples?*

A. Pimples are usually associated with acne, which is a common skin disorder striking especially teenagers. Upon reaching adolescence, more male sex hormones are produced. These in turn excite sebaceous oil glands in the skin to generate more oil. This can lead to the formation of blackheads. Bacteria also play a role. The plugged-up follicle becomes inflamed and acne occurs. If lesions appear they should not be picked at because this can lead to scarring. Usually, acne disappears spontaneously after one to two years and does not remain a permanent problem. In severe cases, treatment is available, which includes hormones and antibiotics. Sunlight is also helpful.

Q. *Why do physicians use the symbol Rx when they write their prescriptions?*

A. There is some controversy as to the exact origin of this symbol, and perhaps all of them we mention here are valid explanations. In Latin, the word *recipi* or *recipere* (which means "take") is abbreviated as Rx. This symbol can also be traced to the sign of Jupiter, as it was found on ancient prescriptions to appeal to the Roman god Jupiter. The crossed R has also been found in ancient medical books wherever the letter *R* occurred. Some argue that the origin is drawn from Egyptian mythology.

The Egyptians believed that there were two brothers named Seth and Horus who ruled as gods over Upper and Lower Egypt. They came into conflict, and in the resulting battle, Seth tore out one of Horus's eyes. But another god, Thoth, was kind enough to heal the eye. Since the eye of Horus consisted of the sun and the moon, and it was the moon eye that was damaged, this explained the phases of the moon; the waning of the moon was the eye being damaged and the waxing, the healing. Thus the eye of Horus became a powerful symbol of healing in the minds of the Egyptians. They also considered Horus the "father of Pharmacy" and wore amulets with this symbol for good health and to protect against illness. When drawn via Egyptian art, the eye of Horus strongly resembles the modern Rx of the physician. It was very early included in magical formulas for healing and has continued to the present day.

Q. *What is a physician listening for when a stethoscope is placed against the chest?*

A. Sounds from within the body offer clues as to whether or not organs are functioning properly. In the beginning, a physician would place his ear against the chest and listen. In 1819, the Frenchman R. T. H. Laënnec constructed a perforated wooden cylinder to hold against the chest. This was the first stethoscope, the term deriving from the Greek word *stethos* meaning "chest." Later, the familiar instrument for listening with both ears came into wide use. Mainly, the physician is listening to the sounds of heart and lungs. For example, if the heart valves are not closing properly, the sound of blood leaking around the valves will manifest as a heart murmur. The stethoscope is also useful for the sounds from women in pregnancy, and intestinal disturbances. The physician calls these diagnostic procedures using the stethoscope *auscultation*.

Q. *What was the reasoning behind the practice of bleeding sick people with leeches during past centuries?*

A. In all ages and most cultures, blood has been intimately associated with the welfare of the person in terms of well-being, strength, character, and other traits. Thus, some drank the blood of the bear or lion to acquire strength, while others avoided touching menstrual blood as it came from women who were thought weak. It is not surprising, then, that when a person was sick, it was thought to be from bad blood. Presumably, if the bad blood were removed, the new blood generated would be free of disease.

Leeches were used to draw blood because they were fairly common, and those attaining a length of four inches or more could remove considerable blood, especially if more than one were applied. Leeching was thought to cure gout, skin diseases, respiratory ailments, and even headaches if applied to the forehead. This is how many people died from loss of blood. One of the most famous victims was George Washington, who caught laryngitis and was bled heavily four times with leeches. That was in 1799.

The leech is actually a worm of which there are three hundred species, both on land and in water. Young leeches can gain entrance to the body through mouth and nose and attach themselves internally, causing anemia and death. The name "leech" is an archaic name for a physician, also meaning to cure or heal.

Q. *What makes my blood type different from someone else's?*

A. Although all human blood looks similar, at the turn of the century four broad groups of blood were recognized at the red blood cell level. Attached to blood cells are molecules called *antigens* and it is the variation among these that establishes your blood type. We have type A with antigen A, type B with antigen B, type AB having both A and B, and type O blood containing neither A nor B antigens. The reason this is important is that if you receive a transfusion of blood of a different type, your antibodies will believe the alien antigens are invaders and attack them, causing clumping or clotting of the blood. Blood has long been associated with magic for good or evil, life and death. Indeed, one reason witches were burned at the stake was in the belief that the witch derived her power from her blood. Fire destroyed the blood.

Q. *In what war did Florence Nightingale become famous for nursing wounded soldiers, and how old was she?*

A. It was during the Crimean War (1854–56) in a military hospital at Scutari, Turkey, that she tended British soldiers.

When she arrived, at age thirty-six, the hospital was rife with filth and sanitation was nonexistent. She completely reorganized all nursing activity and learned the name of every wounded soldier in the hospital. She gained the byname "Lady of the Lamp" from her habit of checking on soldiers in the ward during the night. Nightingale had wanted to be a nurse since her teens, but her wealthy family opposed the idea. Finally, at age thirty, she got the chance to attend a nursing institute in Germany. She graduated, returned to England, and became director of the Institution for the Care of Sick Gentlewomen.

When the war broke out, she volunteered due to the desperate need for nurses. She contracted a fever during the two years of the war and remained an invalid the rest of her life. This did not stop her from founding the Nightingale School for Nurses and being the one to raise nursing to the level of a profession. She persuaded Queen Victoria to appoint a Royal Commission on the Health of the Army, which led to the establishment of the Army Medical School. Nightingale's accomplishments are the more considerable in realizing that she struggled against great prejudice in the man's world of the nineteenth century.

Q. *What disease in history has claimed more lives than any other?*

A. Within recorded history, it appears that the answer is tuberculosis (TB). It is an infectious bacterial disease easily transmitted through the air. It can be transferred by sneezing, coughing, or even standing next to an infected person who clears their throat. It can also be absorbed by drinking unpasteurized milk. The lung is a principle site of the infection although it can attack other organs. In time, lesions appear in the lung, scarring takes place, and the lungs are reduced in their function. The person dies exhausted, trying to breathe. TB was the leading cause of death in the eighteenth and nineteenth centuries, and persists today as a world health problem, especially in the developing countries of Africa and Asia. It is estimated that three million people worldwide die of TB every year. Millions more are already infected. In the United States approximately 20,000 people have the disease in any given year.

Prevention and treatment include quarantine, rest, and a hygienic environment (including pasteurized milk) coupled with administration of drugs. Such drugs, exemplified by streptomycin, are very effective but the bacilli are so durable that they can gain resistance to one or more drugs. A very encouraging development is the complete mapping of the TB genome. This will accelerate research into pinpointing those genes that make these bacilli so virulent, and suggest procedures to render the bacteria harmless to humans. Other strains of TB affect cattle and certain domestic stock such as pigs and chickens.

Q. *Would it be fair to say that the Black Death caused the most fear of any disease?*

A. Yes, it would be not only fair, but also accurate. Consider the fourteenth century in Europe. The plague, brought about by the bacterium *Yersinia pestis*, attacked the population with great swiftness, terrifying symptoms, and quick death. And most sinister of all, no one knew why or how. The population of England was cut in half; hundreds of villages ceased to exist. Kings and queens died of it as well as the lowest farmworker. It is estimated that 25 million people died in Europe, disrupting the entire social order. It took more than a century to recover.

It became known much later that the culprits were rats and fleas, whose bite carried the deadly disease into the bloodstream and to every organ where, after incubation, the victim broke out into raging fever and chills, swelling nodes, weakness, headache, and disorientation.

Death would follow in as little as two days. Once a person had it, others could catch it by the airborne route and this was quickly perceived. People fled into the country, or barricaded themselves in cellars. Yes, it was a frightened Europe in those days. Modern hygiene and rat elimination has checked the plague, but occasional small outbreaks occur.

Q. *Hasn't trichinosis, the disease you get from eating raw pork, been virtually eliminated in this country?*

A. No, it has not. In fact, you are in more danger from trichinosis eating pork products in the United States than perhaps anywhere else in the world. As yet, there is no practical method known for detecting infected pork. Trichinosis is caused by a type of ringworm called *Trichinella*, which is ingested as an encysted form. The cyst is then digested and releases the ringworm which reproduces and spreads through the bloodstream to other parts of the body. This causes fever, soreness, nausea, and diarrhea. However, it is not often fatal. Many people probably have encysted *Trichinella* in their muscle tissues and don't even know it. The solution is to eat only thoroughly cooked pork.

The pig is both loved and hated. Ancient peoples in Crete worshiped the pig. And of course Arabs and Jews will not eat pork. One reason for this abstinence is rooted in the belief that you become what you eat. Yet pigs are intelligent animals. During the times of the Spanish Inquisition, a refusal to eat pork was proof of non-Christian tendency, and some died rather than consume pork. The domestic pig was bred from wild boar stock since ancient times.

Q. *What was Typhoid Mary's real name and what did she do to deserve that name?*

A. Her real name was Mary Mallon. Born about 1870, she worked as a cook in the New York City area around the turn of the century. Although immune to the disease herself, she was an active carrier of typhoid fever. At that time, an untreated victim died within twenty-one days of the onset and there were no antibiotics to counter typhoid. It was known, however, that a carrier could infect other people because the bacterium, *Salmonella typhi*, entered the body through the mouth by eating and drinking. Mary was a food handler and knew this.

An outbreak of typhoid in 1904 on Long Island was traced directly to her by health authorities. She escaped. Three years later another outbreak occurred and she was caught after a long chase. She was confined until 1910 and then released under the promise that she would not work as a cook anymore. In 1914, an epidemic broke out in Manhattan and it was Mary again. She escaped but was caught. This time she was held at the isolation center on North Brother Island until her death in 1938. It is a sad story.

Typhoid bacteria are associated with polluted water and sewage. In unsanitary areas, typhoid may be expected. An unsuspecting oyster in a polluted bay may be the Typhoid Mary of today. Typhoid fever is not typhus. They are two different diseases.

Q. *How does a person suffer injury or death from an electric shock?*

A. It is due to the passage of a strong electric current through the body, especially if the greatest current density traverses the heart. This causes the heart to undergo paralysis, twitch, and become irregular, a condition known as *ventricular fibrillation*. The voltage does not seem as critical as the amount and duration of the current. Most deaths from electric shock are caused by alternating current (AC). Most shocks are the result of contact of hand or arm with the electric source such as a power line. In this case, the current pathway is through the heart. If contact is made with the head, the current passes through the brain, possibly causing damage.

Physicians are not unmindful of shock risks even in hospitals where so much sophisticated equipment is used and is often attached to the patient. While we think of electroshock treatments as a rather modern development, early Greeks and Romans were using the torpedo fish, or electric ray, which carries an electric charge, to try to cure gout and other ailments.

Q. *Do natural mineral crystals have the power to cure mental or physical illnesses?*

A. No. Crystals have an outward symmetry reflecting an internal ordering of atoms produced by the slow growth of the crystal in a fluid. They are as natural as frost on the window in the winter and have no magical properties nor are they surrounded by electromagnetic or other force fields. Crystals are relatively rare in nature and it is understandable that people of all ages in the past believed them to possess some supernatural power. Actually, the crystal ball the fortune-teller uses isn't a crystal at all, but rather a mass of homogeneous glass with only random ordering of atoms of silicon and oxygen. Claims that crystals will solve your problems with some kind of quick fix should be looked upon skeptically. We might place magnets in the same category. People already live within Earth's magnetic field without noticing any effect, and there is no credible experimental evidence that wearing a magnet on your wrist or in your shoes will cure anything. If you are sick, see a doctor, not a crystal gazer.

Q. *I read about a ninety-five-year-old man who has had the hiccups for sixty-seven years. What would cause something like this?*

A. While hiccups (hiccoughs) are common, the case you describe is unusual. Hiccups result from a spasm of the diaphragm, which in turn shuts off air inflow through the upper part of the larynx between the vocal cords. There is a wide spectrum of causes including stomach disorders, pneumonia, kidney problems, and even pregnancy. Excessive alcohol is also a cause as in the caricature of the hiccuping drunk under the streetlight. Psychological causes may also be involved as in the case of some disturbed children.

Time-honored cures include holding your breath, drinking a glass of water, and breathing into a paper bag, all of which work most of the time. The reason is that low carbon dioxide in the blood encourages hiccups while high carbon dioxide suppresses them. The above tactics increase carbon dioxide. If these methods fail, a physician can prescribe drugs, but even medical science admits that not all cases are curable, such as the ninety-five-year-old man. Men get hiccups more than women do.

Q. *My friend is being treated by a homeopath. Just what branch of medicine is this?*

A. Strictly speaking, it is not a branch of orthodox medicine, nor does organized medicine hold it in very high regard. It is not discussed in the authoritative Merck Manual, for example.

Homeopathy, or homoeopathy, was founded in the nineteenth century by the German physician Samuel Hahnemann. His basic idea was to administer a drug producing the same effects as a given complaint, but only in very small doses. Thus, if you have a headache, take something in small amounts that also causes headaches. These small amounts of a drug would act as a trigger mechanism and set the body to healing itself. Over the years homeopaths have gathered up to one thousand highly diluted drugs that can be matched with about any ailment you can imagine. However, the dilution is so great that the drugs applied can be regarded as chemically inert.

A major criticism from the medical community is that the body of experimental evidence is insufficient for homeopathy to have credibility. In the past several years, charlatans have marketed useless goods by direct mail or in health-food stores claiming they are homeopathic.

Q. *Was the inventor of the Braille system of writing for the blind also blind himself?*

A. Yes, but not born so. The Frenchman Louis Braille suffered an accident at age three, which rendered him permanently blind. His system came about while he was still a young student at the National Institute for Blind Children in Paris. He got the idea from a method called *night writing*, conceived by Captain Charles Barbier, who saw in it the possibility of use under battlefield conditions. Braille was only fifteen years old when he modified Barbier's method into what is now the Braille system.

In Braille, there are six basic embossments sensible to the light touch of the fingers. These are arranged in various positions to form sixty-three characters. Braille users also have a slate to write on as it is called, and Braille typewriters are available. Braille died in 1852 at the age of forty-three of tuberculosis.

Q. *Aren't reports of persons with enormous appetites, such as Diamond Jim Brady, highly exaggerated?*

A. Perhaps not. Brady was a generous millionaire who dined frequently in New York City restaurants during the last century. Many witnessed a typical breakfast: eggs, bacon, bread, a gallon of fruit juice, a load of pancakes, fried potatoes, and steak—enough for five men. He got his nickname for his penchant of wearing diamonds routinely. Brady bestowed considerable sums of money on various charities.

A rival for Brady as eating champion might have been King Louis XIV of France who consumed enormous amounts of food at one sitting. There was a reason: after his death, it was found that Louis had a stomach twice the normal size. Some people are unable to assimilate the full nutritional value of food and thus are constantly eating. Those with a surgically shortened intestinal tract may eat more. It is along the tract that nutrition is absorbed. Still others may eat so fast that the hunger mechanism isn't turned off until they have overeaten.

Q. *Are there actual cases where a person's hair has turned white overnight because of some shock?*

A. There is no scientific basis for this belief. Once hair extrudes from the hair follicle in the scalp, it is no longer living tissue. A hair is only alive at the root where the substance *keratin* makes up the hair and a pigment called *melanin* determines its color. Thus, there is no human emotional reaction or trauma that can have an effect on already dead material. Under prolonged stress, it is possible for melanin to decrease and this can result in a gradual graying of the hair. To say it happens overnight is to exaggerate and some people like to be dramatic.

Melanin is also an eye pigment that protects the eyes from harmful radiation. It is also the pigment that browns the skin when you get a suntan. This safeguards the skin from the sun's rays. Albinos have no melanin.

Both hair and nails can be thought of as extensions of the skin, mainly comprised of keratin. Incidentally, the average scalp contains 125,000 hairs, unless you are bald.

Q. What good is your spleen and can a person live without it?

A. It performs a number of useful functions in resisting infection and maintaining the quality of the blood. But, yes, you can live without it. However, it is quite important in infants under age two in preventing infection and producing red blood cells. As one gets older, the spleen's importance diminishes. The spleen acts as a filtration system for the blood. For example, red blood cells have a life span of 120 days. After this, the spleen removes the used-up cells but recycles the iron in these cells for other purposes. In some lower animals, the spleen acts as a reservoir for extra blood, which can be dumped into the bloodstream by spleen contraction if the animal is faced with a fight-or-flight situation. Enlargement of the spleen is called *splenomegaly* and may suggest disorders elsewhere in the body. If damaged, one may have a *splenectomy*, where the spleen is surgically removed. However, people who have had this surgery can lead normal lives and have normal life spans.

Q. Has anyone been accidentally buried alive?

A. In earlier times, yes. Bodies exhumed from the last century and earlier bear the unmistakable signs of the person having thrashed around trying to get out. It is a gruesome thought, but this could happen before there was a medical understanding of people lapsing into a coma or other condition resembling death. Today, with the widespread practice of embalming and superior medical knowledge to distinguish death, it is impossible or at least highly unlikely. Perhaps some ghost stories originated from the fact that individuals were interred alive in above-ground crypts and were able to escape from their coffins and wander around. We might add that the notion that hair and fingernails continue to grow after death is a myth. The apparent growth results when the surrounding tissue shrinks.

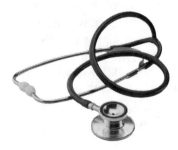

PART 4

THE EMERGENCE OF HUMANITY

11

PRECIVILIZATION
Hunters and Gatherers

Q. *According to scientists, how long have humans been on Earth?*

A. We do not know exactly, but anthropologists are working on it. The evolution involved can be likened to a tree with a main trunk and many limbs and branches. Africa seems to be the favored place where humans first emerged, but fossil *hominids*, the superfamily that includes apes and humans, have also been discovered in Asia, Europe, and the Middle East. Hundreds of fossil hominid bones and teeth have been found in African locations such as Olduvai Gorge. Stream erosion at this important site has exposed successive layers of sedimentary rocks, establishing a chronology of occupation in this region. Among the fossils found are those of the *australopithecines* which represent an important link to the genus *Homo* and which flourished from about five to two million years ago. From between 2.5 and 1.5 million years ago we find evidence of *H. habilis*, and from 1.6 million years, *H. erectus*, both representing evolutionary steps toward our own species, *H. sapiens*, but the relationships are not clear. Many scientists favor the idea that *H. sapiens* evolved from *H. erectus* about 200,000 to 300,000 years ago.

Q. *I see references to the Stone Age, the Paleolithic, the Bronze Age, and so on. Can you sort these out for me?*

A. Once humans appeared on Earth, they began to make tools, such as those used for cutting, grinding, and scraping. Also, hunting tools such as projectile points were developed. Archaeologists have established a chronology based on toolmaking and the materials used. The Stone Age was a period of making and using stone implements. It can be subdivided into the old Stone Age (the Paleolithic) and the new Stone Age (the Neolithic). The Paleolithic extended up until about 30,000 years ago and involved the making of crude stone tools. From 30,000 to 10,000 B.C.E. was the Neolithic, during which more sophisticated stone tools were made; this was also the time of cave paintings and elaborate burials. Subsequently, the use of metals rather than stones ushered in the Bronze Age about 12,000 years ago, and after that came the Iron Age. In the geologic time scale, this entire period is known as the Pleistocene (meaning "most recent") and was of 1.6 million years duration. This epoch was marked by four major advances and retreats of thick ice sheets. A time term popular among geologists is *Holocene* (meaning "entirely recent") to indicate the interval from the retreat of the last ice sheet 10,000 to 12,000 years ago, to the present. Still, other geologists claim we are yet in the Pleistocene and the ice sheets may advance again.

Q. *What was the closest prehuman to our own species?*

A. Neanderthal Man used to be a good candidate but recent thought is to classify the Neanderthals as human. Two other possibilities would be *H. habilis* and *H. erectus*, but these species are regarded by many as basically human. A problem here is that at the time humanoids appeared there was so much evolutionary raw data available that it created confusing variations. Adding to the problem is the fragmentary nature and poor conditions of preservation of the fossils, which must be painstakingly assembled by anthropologists who often need to infer their conclusions. Even the skeleton of the well-known "Lucy," which so thrilled the scientific community, is only 40 percent complete. Thus, the answer is that we are not sure. It is possible that among the highest of the australopithecine group, there is a creature that would qualify. Lucy herself remains a viable candidate, and if so, then the closest prehumans would extend back in time three million years.

Q. *What did Australopithecus look like?*

A. It was a small, apelike creature, about five feet tall, weighing between 75 and 130 pounds. It had a cranial (brain) capacity of perhaps 430 cubic centimeters, which, as a ratio to body weight, placed it intermediate between apes and humans. *Australopithecus* was first discovered as cave remains in South Africa in 1924 by Raymond Dart. Three species were established: *A. africanus*, *A. afarensis*, and *A. robustus*. The limbs are adapted for climbing, but bone structure suggests an ability to walk upright. The teeth are a mix of apelike and human characteristics. Australopithecus possessed hands rather than claws, and would have been capable of making and using primitive tools. Associated fossils and artifacts suggest they hunted or scavenged the remains of antelope and baboons, probably in groups although this is uncertain.

Q. *Where does "Lucy" fit into the scheme of human development?*

A. Lucy, the popular nickname given to this small female, was an australopithecine of the species *afarensis*. She was found to be three million years old by the argon gas method. She was three and a half feet in height and appeared to have died a natural death, and then was progressively buried under volcanic mud, ash, and other fine sediment. The significance of Lucy was that she was bipedal, according to knee structure, and could walk upright. Her knee bones could "lock" into place and sustain locomotion. Dr. Mary Leakey confirmed this by finding footprints in hardened volcanic ash of similar age to Lucy. The prints showed a big toe not unlike a human and also an arch, certainly not an ape trait. Leakey observed that more than one individual made the footprints. It may be inferred that such prehumans, in walking, could cover a larger foraging area, and had hands free to carry food, tools, and weapons. Such advantages represent a great stride forward (no pun intended!) in evolution toward the genus *Homo*.

Q. *When were the first tools made, and who fashioned and used them?*

A. The first deliberately made tools were of stone and some bone. It is also possible, even probable, that many early tools were made of wood, but have not survived. A case can be made that the oldest tools come from the Hadar region of Ethiopia, in Africa, and have been dated at two and a half million years. These artifacts have been found there by the hundreds. That date would place toolmaking at a time when australopithecines were widespread and would show that they had the intelligence and ability to make useful cutting and striking tools and even small axes. At the same time, *Homo habilis* had arrived on the scene and some of these ancient tools might be ascribed to their making. It is likely that these groups subsisted chiefly on a vegetarian diet. It was easier and safer to do so. However, during dry seasons when edible plants were scarce, australopithecines and *H. habilis* may have made cutting and scraping tools. These would have helped to obtain the last meat remnants of a predator's kill, or to smash open bones to eat the nourishing bone marrow. Many animals, including birds and small mammals, might use a casual, ready-made tool such as a stick or a rock, but a tool conceived and designed with a particular purpose in mind requires intelligence and skill.

Q. *What was the effect of climate on human advancement? It must have been hard to survive during the ice ages.*

A. Any large-scale climatic change would have a profound effect on behavior and survival techniques of not only hominids, but other creatures as well. There is the ripple effect of climatic change causing a different array of vegetation to form and some to die, and in turn, the distribution of animals within the food chain. Such occurrences can lead to extinctions, as we think happened to mammoths and mastodons. For others, it can be an opportunity. During the ice ages of the Pleistocene epoch, the colder weather may have accelerated, for early humans like *H. erectus*, the discovery and use of fire for warmth, light, and cooking. Fire also could be seen as a weapon to ward off predators. The earliest clothing would have been animal pelts. Hunting methods may have been perfected for cold-weather animals such as reindeer and musk oxen, and at the same time facilitated cooperative behavior. Cave living would have been favored for protection and safety, and, in the forced leisure time imposed by frigid conditions, crafts and other creative pursuits could develop. In contrast, a benevolent climate with abundant food and few predators may have diminished the survival capabilities of many animal groups, including humans.

Q. *Flies and mosquitoes drive me crazy, but we have insect sprays. What about primitive man? Did they have some special way to cope with this problem?*

A. Anthropologists have no particular answer to this question except to say that early humans enjoyed the attention of these insects no less than we do today. Even now, in less advanced countries smoke pots are still used to drive insects away. Consider also that we humans seem to be able to get used to anything we regard as a part of our lives. People in less advanced societies might be covered with various kinds of bugs around the head and body, yet pay little attention to them. Yet those of us who do not usually experience this kind of attention and are not used to it can spend an entire night chasing one mosquito around the bedroom. Primitive humans may have used smoky fires as well, or rubbed animal fat on the skin as protection. The animal pelts worn then may also have helped. It is a case of adaptability, whether pleasant or unpleasant, and our species has proven very good at that.

Q. *Is there a definite progression of intelligence in the evolution leading to* **Homo sapiens***?*

A. Yes, and it has a number of ramifications. If we take it that increasing brain size and its ratio to body dimensions means increasing intelligence, then there is a fairly clear progression. The australopithecines are at the low end of the scale with a capacity of about 450 to 500 cc (cubic centimeters) followed by *H. habilis* at 700 to 800 cc. *H. erectus* had a capacity ranging from 1,000 cc to as high as 1,200 cc, while a typical *H. sapiens* measures 1,450 cc cranial capacity. The *Homo* brain could not increase thusly without the head also becoming larger. Such large heads would have difficulty passing through the birth canal.

Some scientists visualize that, in order to overcome this physiological impasse, births started to take place before the head of the unborn infant became too large. In contrast, most mammalians are born with the brain fully developed and are fully functional within a short time. In the case of humans, the brain would not have achieved this level and would continue to develop, and the head size increase, after birth. It follows that the new offspring would be virtually helpless, requiring parental nurturing and protection for a more sustained period of time. A beneficial side effect in terms of mutual survival would be a social one. A clan or group would be more cohesive, staying together longer with increasing mutual cooperation and a heightened sense of consciousness-of-kind. We might then imagine a more efficient pooling of tribal resources that otherwise might not have occurred had offspring been more independent and precocious. It seems like a plausible scenario, given what we know of human birth and infant care in today's world.

Q. *Why has the Neanderthal been depicted as having low intelligence and only able to grunt and snarl?*

A. Neanderthals have been much maligned. The popular concept of a shuffling, stooping creature is due to the faulty analysis of skeletal material—the individual studied had suffered from arthritis. Neanderthals were not that different physically from modern humans, but stockier and more muscular, with a strongly projected face. We do not know the extent of their power of speed, but they were able to communicate. The Neanderthals were hardy, aggressive, and survival-oriented. They were able to prosper for at least 200,000 years across Europe, Africa, the Middle East, and elsewhere, and braved the ice ages. This is successful evolution. It is unlikely that they were vicious beings intent on wiping out the increasing numbers of Cro-Magnon humans in their own territory.

Q. *What is the current thinking by scientists regarding the origin and fate of the Neanderthals?*

A. After several decades of believing the Neanderthals to have been subhuman cave dwellers, scientists have greatly revised their opinion, persuaded in large measure by the wealth of skeletal and associated artifacts recovered at the many Neanderthal sites across Europe, Africa, and western Asia. This revision of opinion is reflected in the assignment of the Neanderthals to our own species as *Homo sapiens neanderthalensis* and with evidence to justify it.

Although the precise origins of the Neanderthal remain shadowy, we can see in later bone remains of *H. erectus* the appearance of some Neanderthal characteristics suggesting that the evolution of the Neanderthals paralleled that of our own species. While their heads were shaped differently from our own, their brains were generally as large as those of modern humans. They lived in small groups, mostly as hunters, using spears, of small game, having knowledge of fire as their excavated sites indicate. The tools they made ranged from crude flakes to more sophisticated stone tools, and they even hung objects of adornment around their necks. Significantly, they buried their dead, with indications of grave goods. The Neanderthals may well have been the first creatures on Earth to have some concept of an afterlife.

The Neanderthals had hard lives. Their bones show fractures, breaks, and dietary deficiency in the skeletons of three out of four individuals. Few lived beyond the age of forty. There is some evidence emerging that on occasion they practiced cannibalism, or were forced to it. Their retreat into extinction coincided with the spread of the Cro-Magnon, who in all likelihood assumed control over a common habitat. Since the two groups lived at the same time for several millennia, it is probable they interbred and the Neanderthals were absorbed.

A recent discovery of a four-year-old child's grave in Portugal in 1998 may shed some light on the fate of the Neanderthals. The scientists determined the date of the burial at around 25,000 years ago, some 5,000 years after Neanderthals supposedly disappeared. The bones of the child have characteristics of both the Neanderthal and early modern humans, which suggests that the Neanderthals had been absorbed through interbreeding.

The midfacial prognathism, a protrusion of the nose and jaw, of some western Europeans also gives some support to the idea of interbreeding. If you are a descendent of a west European, there is a chance that both Cro-Magnon and Neanderthal blood runs in your veins. On the other hand, studies of DNA taken from a Neanderthal bone suggest that there is no relation of Neanderthals to modern humans. Investigations continue on this controversial subject but we do know that sometime between 50,000 and 30,000 years ago, the Neanderthals disappeared altogether as a distinctive subspecies of *H. sapiens*.

Q. *What were the main differences between these two early humans, H. habilis and H. erectus? Did they look like us?*

A. The richly fossiliferous Olduvai Gorge was again the scene of discovery. In 1959, skull material was unearthed that was clearly different from earlier australopithecines and bore characteristics sufficient to place this creature in the genus *Homo*. It was given the species name *habilis*, meaning "handy man." Habilis had a larger brain and smaller teeth, with a more rounded skull than its predecessors, but still retained enough apelike features to set it apart from more modern humans. Several other sites yielded individuals of this species, including the important site of Koobi Fora. Richard Leakey opined that *habilis* represented the earliest human known, having dated the skeletal material at very close to two million years BP (before present).

During the late nineteenth and into the twentieth century, a large number of skulls, mandibles, and other bones were discovered in China, Indonesia, and Africa that seemed related but featured a wide range of morphologic variability. One trait in common was a brain capacity with an average of 900 to 1,100 cc, larger than *habilis*, but generally less than *H. sapiens*. This new human creature, dubbed *H. erectus*, was dated as early as 1.6 million years and appeared to be an evolutionary descendent of *H. habilis*. However, these people made better stone tools, including sharp-edged choppers and bifacial hand axes. They were more along the line of true hunters than *habilis*, who scavenged perhaps more than hunted, and they enjoyed a wider variety of diet including fruits, roots, and nuts. While their place in the *Homo* lineage is established, there is considerable debate among scientists as to their exact connection to archaic humans who belonged to our own species. Both of these groups show an evolution toward a more modern human appearance, especially in nose and chin, but still would have looked unmistakably "primitive" to us. The fossil record does not reveal such characteristics as color of eyes, color and texture of skin, hairiness, or subtle behavioral traits. We remain ignorant also to what extent these groups possessed linguistic skills, or the throat structure to articulate them.

Q. *When did these Cro-Magnon people appear, did they usually live in caves, and were they as smart as we are?*

A. Cro-Magnon is the name of an archaeological site in France, discovered in 1868, which contained the skeletal remains of ten fully human individuals dating 10,000 to 35,000 years ago. This site was a rock shelter, although others of this time did live in caves or in the open if the climate was benevolent. However, probably most of the time it was cold as the effects of the ice ages were still being felt. Nonetheless, they were widespread geographically and in habitat. These people fulfilled in general the definition of *H. sapiens* in brain size, vertical forehead and rounded back of the skull, a distinct but pointed chin, and a dental arcade that would be familiar to a modern dentist. It is likely they had the same native intelligence as we do, but of course not the knowledge that civilization built and built upon. They were the tallest of the genus *Homo* in those days, some perhaps being six-footers, and the skeleton was fully adapted to the ambulation and gait of modern humans. Many scientists believe Cro-Magnon evolved from some subspecies of *H. erectus*, or a more archaic sapient type extending back 130,000 to 150,000 years.

The cultural advancements of the Cro-Magnon are of interest. They built small huts or lean-tos, and fashioned a variety of useful tools (including many of bone) such as scrapers, smoothers, awls, and spearpoints with which they successfully hunted large animals including mastodons. They can be considered the first true artisans. They made small sculptures, engraved geometric designs, and fashioned adornments, and were the artists responsible for the famous cave paintings in France and Spain. Their encampments seem to have been more permanent rather than seasonal abodes, indicating a drift toward more settled communities that foreshadowed the first civilizations. They may have believed in an afterlife and had a concept of religion as suggested by some of the artifacts. They must have had some linguistic communication but no written language. Further archaeological finds may bring to light additional knowledge of their lifestyles.

Q. *Are humans the only creatures that make war and kill members of their own species?*

A. When its own life is threatened, or that of its offspring, an animal, regardless of species as far as we know, will attack and kill its own species. Humans are certainly no exception to this. Among the mammals, a male may attempt to kill the young of his tribe or group because they appear to him as a future threat to the adult male's superiority. The mother will defend the offspring to the death. This is fairly obvious. With humans, it may be less obvious when two countries go to war and soldiers die. Yet on a more sophisticated level, the war is often fought because a threat to survival is perceived.

It may actually be that our aggressive traits were instrumental in our survival. Consider other animals. Many species of birds, reptiles, mammals, and amphibians show aggressive traits. This is seen in the animal's staking out an area or territory containing the resources needed for survival. This territorial instinct intensifies during both mating season and the rearing of the young. Some animals stake out large areas and offer combat to those that invade, sometimes to the death. Some, such as a bird, may only consider the nest itself as the defended territory. Where animals are not aggressive, we may see extinction. An example is the passenger pigeon, which offered no resistance to egg hunters destroying the nest, and thus was wiped out in the twentieth century. Aggression in humans can be seen as normal behavior. Our problem is we keep dreaming up more and more efficient ways of expressing this aggression.

Q. *What was the world's greatest archaeological fraud?*

A. Some would say it was the famous Piltdown forgery. *Piltdown Man*, the phony fossil, was "discovered" in a gravel formation in southern England in 1912. The unknown person or persons who combined an ape jaw (an orangutan) with skull fragments of a modern human did it with such great skill that scientists argued over Piltdown for about forty years. The forger had filed the teeth in the ape jaw to give them the appearance of a human wear pattern. The bone had been chemically treated to simulate long burial and antiquity. Such deft touches indicated that perhaps a forger from the archaeological ranks was the culprit. Many anatomists of the day accepted Piltdown without question as a primitive form of a human but chemical investigations in the 1950s showed the cranium and jaw to be of different ages. The damage done to scientists seeking knowledge of human ancestry was considerable, as it set them upon a false trail of assumptions over that forty-year period.

Once the forgery came to light, practically everybody who had any connection with the original discovery became a suspect. Charles Dawson, a lawyer who found the fossil in the first place, was a primary suspect. He was thought to be a collaborator with a more technically trained conspirator. Suspicion centered on Arthur Smith Woodward who was with the British Museum and would have had access to the material of the phony fossil. He also had a motive in that he aspired to be the director of the museum but had been turned down more than once. Such a great discovery would speed his professional advancement. Another suspect was the creator of Sherlock Holmes, Sir Arthur Conan Doyle. He lived near the site and was an ardent antievolutionist. Plausible arguments were marshaled against these major suspects. However, a "smoking gun" emerged in 1975 with the finding of a steamer trunk belonging to Martin Hinton, a museum volunteer who had a grudge against Woodward. The trunk contained bones and the chemicals to artificially age them identical to those that had been used in the actual hoax. Unless further evidence to the contrary is found, Hinton stands as the culprit in the Piltdown hoax—probably a mystery Doyle would have liked Sherlock Holmes to solve.

12

EARLY CIVILIZATIONS
Getting Organized

Q. *What is it that characterizes any civilization?*

A. There are a dozen ways to define civilization, but at its most fundamental level, a civilization is predicated upon cooperation to increase and maintain food production. Everybody has to eat. With the end of the ice ages 10,000 to 12,000 years ago, there were groups of *Homo sapiens sapiens* scattered throughout much of Africa, Europe, Asia, and the Mediterranean area. These small bands of twenty-five to fifty individuals were much like seeds sown across the soils of a world of increasingly mild and hospitable livability. The climate was growing warmer and wetter. When they occupied a favorable area of abundant game and vegetable victuals, there was a tendency to linger longer. These communities grew in numbers and in their capability to exploit food resources. Thus arose the very early cultures of Quadan in the Nile Valley and Natufian in the Levant (Syria, Lebanon, and Israel) as the ice ages ended. The nomadic life was slowly abandoned in favor of settled locales where several hundred might prosper. To increase efficiency, we can visualize a division of labor, more permanent dwellings, better linguistic communication, leadership and new ideas, and eventually the evolution of the first cities. It took several thousand years for this to come about, but by 7000 B.C.E., village-farming communities in the Middle East were well established. It would be from these small wellsprings that the great ancient civilizations of the Sumerians, Egyptians, and Babylonians would emerge.

Q. *What wild animals did early man domesticate?*

A. It may have been the dog because while early humans were still in a primarily hunting stage, the dog would be useful as a hunting companion and as a guard at night by encampments. The dog could also be eaten. Still, this is disputed by some. A case can be made that a herd animal such as the goat, sheep, or pig would be easier to tame and conveniently provide meat, milk, wool, and skins without the necessity of hunting them. The finding of bones of these animals at human occupational sites dating to 9000 B.C.E. seems to bear this out. As agriculture advanced, work animals like oxen, asses, camels, and cattle would be domesticated as needs arose. The domestication of some wild animals, such as the hyena, was less successful. Horses would come much later in the steppes of Asia. The larger and stronger herd animals, including people, were recognized as exploitable for transportation, and from this undoubtedly came the invention of the wheel, which was probably invented several times over in various cultures around the world.

Q. *What is the earliest evidence of agriculture?*

A. Its earliest beginnings probably occurred when humans discovered that certain wild cereals, such as wheat and barley, were good to eat and could be collected from large stands of this resource. Archaeologists have found mortars, grinders, and sickles associated with other artifacts of the Natufian culture at numerous sites in the Levant. However, it is known that such discoveries were being made elsewhere in the world by other cultures and at differing times within the range 12,000 to 9,000 BP and perhaps even earlier. It would be apparent that seeds of these grains, fallen to the ground, eventually produced more of the same. Therefore, you could plant the grain you wanted, in whatever quantity, and at convenient places. Another step forward was the recognition that the grain grew better in some places such as flood plains next to rivers. Experiments with hybrids and the use of irrigation would follow. It was no accident that the great civilizations arose along major rivers as the Nile, and the Mesopotamian Tigris and Euphrates rivers of the Fertile Crescent. Those early cultures already practicing domestication and herding of various animals would be favored in faster development because the animals could be employed in planting, tilling, and harvesting of crops. In this way, domestication of animals and plants for deliberate aims formed the foundation of ancient civilizations.

Q. *I was eating a slice of bread and wondered who first thought of it—and I'm not even sure how you make bread.*

A. Consider that more than 20 billion pounds of bread are consumed in the United States alone each year. Quite a number of the consumers probably don't know how it is made either. Archaeological evidence shows that bread was made at least 10,000 years ago in the region that is today Switzerland. This was not our modern bread, but rather ground acorn mixed with water and heated to form cakes. Later, bread was made with ground cereals such as wheat, rye, barley, and corn. The ancient Egyptians were apparently the first to make white bread from wheat and "raise" it by using yeast. Archaeologists have found ancient ovens where the bread was baked. With regard to bread's "invention," accident likely played a major role, as it did in fire and cooking, the wheel, and the bow and arrow. The bread that is eaten today was a luxury reserved for royalty as late as the seventeenth century. Commoners ate the flat, rather tasteless bread similar to that made perhaps thousands of years before.

Q. *Archaeologists often have to dig below the ground to find ancient ruined buildings. How did these places get buried that were at one time at the surface?*

A. Archaeologists have to excavate because the workings of nature help to conceal ruins, fossils, and other relics. For example, consider an abandoned barn. Collapsed from neglect, the wood begins to decay. Wind and rain bring in soil whereupon weeds and trees take root. The plants discard leaves and die. This organic debris piles up, a new soil may form, and over time the barn becomes completely buried. Other buildings may be burned, torn down, or partly destroyed (in war, for example) and new structures built on top of the old. The old land surface then becomes artificially buried. A case in point are buildings of the Aztec civilization found below the streets of Mexico City.

Q. *Which did humankind domesticate first, dogs or cats?*

A. Skeletons of dogs have been found in the remains of human habitations dating back at least 10,000 years. Dog experts believe our modern dogs are derived from the gray wolf, which roamed all over Europe in primitive times. We can imagine as the starting point a hunter finding a litter of orphaned puppies and taking them home. There seems little doubt that the dog came first. It may also have been the first animal that humans attempted to breed for enhancing desirable traits such as better sight and smell, strength, and swiftness. The affection and bonding that developed between man and dog is reflected in the finding of graves of chiefs and other tribal leaders buried with their dogs thousands of years ago.

Members of the cat family were well established three to four million years ago during the Pliocene epoch. Cats, including the Abyssinian and Persian types, appear to have been domesticated much later than dogs. They may have originated in Africa, but were known and domesticated by many societies. We do know that cats had become pets of the ancient Egyptians from mural scenes of daily life, who mummified them, and even had cat cemeteries 5,000 years ago. Indeed, the Egyptians worshiped the cat as a sacred animal. The cat was not only an affectionate pet, but kept down the rodent population in and around granaries. Persons in ancient Egypt who deliberately harmed or killed a cat were themselves put to death. If you own a cat, it may have Egyptian ancestry.

Q. *When did humans first start using pottery and what significance does it have?*

A. The basic idea that pliable clay could be molded by hand and then hardened by heat to create a variety of objects was discovered about 28,000 years ago by some early peoples in Europe. Perhaps it was another accident, such as a lump of soft clay falling into a campfire. In any event, the clay was used to make figurines or other decorative objects, but seemed to serve no utilitarian purpose. Then there was a long period of time without ceramic material until its rediscovery near the end of the ice ages when a blossoming of pottery making took place in many widespread cultures. Pots and other vessels could be used to carry water and store things in. The earliest pots were made by the Japanese, and their long-dead finger and nail imprints can still be seen on the pottery.

To an archaeologist, pottery can sometimes be a discovery more important than that of a chest of gold coins. Pottery has been a near-ubiquitous discovery at archaeological sites for more than 10,000 years. This means that it is also very durable and chemically stable. Pots break, of course, but the pieces—called *sherds* by archaeologists—do not disintegrate readily. Before firing, the clay was shaped and decorated in a virtually unlimited number of forms and styles, and temper such as sand or straw added to bind it together. A very important consideration is that pottery styles changed through time, providing a means to identify the culture and estimate age. Styles are datable in the same way that automobiles can be dated according to their design. Although we usually think of pots, fired clay was also used for jewelry, lamps, brick construction material, and as tablets for the earliest form of writing. Thus, pottery is for the archaeologist one of the single most important remains from the past.

Q. *Today, archaeologists use ancient pottery as a main evidence of early peoples. Thousands of years from now, what do you think they will look for from our civilization?*

A. Our civilization is far more widespread and complex than earlier ones and so the numbers and types of artifacts would be in the thousands. Also, opportunities for preservation would be greater because a significant part of our civilization is underground and protected from destructive elements. We may assume that a few bank vaults, the below ground levels of large buildings, and subway systems might be preserved. With time, objects made of iron would crumble away but many things would survive, such as glass, ceramics, and bricks, as well as reinforced concrete, some plastics, and wood. Tapes, compact disks, and film might not last, but the written word is so extensive that we imagine future archaeologists would soon translate our languages. Another factor involved is that these archaeologists may well have very sophisticated excavation and preserving techniques that reveal more than present methods. If it will be an atomic catastrophe that destroys us, probably much less would remain.

Q. *Languages are so complicated, how could any primitive society manage more than some simple words and gestures?*

A. Perhaps at first that is all they did. But if there is cultural continuity, language is something that can be built upon. Extensive use of language is uniquely related to *Homo sapiens sapiens* and one of the most difficult to study because, unlike tools and pottery, a noise from the throat evaporates in the air and leaves no trace behind. Humans, as well as most other animals, make sounds expressing pain, danger, and fear. A person may accidentally touch a flame and scream "ouch!" This is an emotional vocal response to a situation shared with many animals. On the other hand, a person can gaze into the fire and say, "Fire is hot." This is discursive language tied to a thought.

Humans, with many thoughts, both concrete and abstract, and the means of articulating sounds to represent them, possessed early on the potential for linguistic growth. We might discern that the first rudimentary languages were strongly linked to survival, and the first words were sounds representing such items as food, game, hunger, water, grain, danger, come here, sleep, eat, run, and fight. Many gestures or other bodily motions found vocal expression. With greater survival ability and increased leisure, more abstract thoughts such as gods, beauty, life, and truth found vocal expression. Mostly, we can only surmise linguistic development until this form of communication crystallized into written language, which can be studied and analyzed.

Q. *What were the earliest attempts at a written language?*

A. The pictograph made sense and was in use 5,000 years ago in several cultures, including those of the Egyptians and the Sumerians. The first pictographs were drawings in soft clay of the object or idea one had in mind, such as a bird: a bird would be drawn. The Egyptians used a wavy line to denote water. Placed in proper sequence, ideas were formed. To make these representations more permanent, the clay tablets could be baked in the sun or heated in ovens. As city-states developed in Sumeria along the Tigris and Euphrates Rivers, local kings and leaders used pictographs as seals or seal cylinders to represent their names, proprietorship, and accomplishments. With the expansion of trade and commerce among Sumerian cities, the necessity grew for a lasting record of business and government transactions in some written form. Thus, economics played a major role in the development of writing and literacy. A major drawback to Sumerian pictographs was that many had no correspondence with the spoken language. It could not be read out loud and understood.

It is amusing that modern businesses today use pictographs and call them logos to symbolize themselves. The study of pictographs is called *logography*. Likewise, in our computer-oriented societies, a typical computer is replete with icons, also pictographs. An advantage of the pictograph was its ability to transcend spoken language and be understood by all as we see in the icons gracing the doors of public restrooms today. Yet, these picture languages of the past were cumbersome and limited in the degree of thought that could be conveyed. In time, they would be replaced by more efficient systems using an alphabet.

Q. *Who were these mysterious Akkadians that are not mentioned much in popular history books?*

A. It can be said at the outset that the history of the Fertile Crescent is very complex and incompletely known by archaeologists and historians. With the rise of city-states there was interplay among various Semitic tribes—the Sumerians, Assyrians, Babylonians, Elamites, and Akkadians. The third millennium B.C.E. seemed to be a time of endless warfare among them all, yet at the same time was a golden age of advancement in art, architecture, writing, and science, particularly in Sumer, which justifiably might be considered the place civilization began.

As to the Akkadians, they were semitic communal groups in villages and towns in the region north of Sumer. As trade routes developed, villages and caravans suffered from raids and pillage from outlying lawless tribes. There arose in Akkad about 2350 B.C.E. a wise ruler named Sargon who was also a warrior. He formed armies to destroy lawlessness and bring security to the region. Succeeding beyond all expectations, he went on to conquer all the major cities of Sumer as far south as Ur and Uruk. For the next 200 years, the Akkadians held military dominion over the Fertile Crescent through Sargon's sons and their descendants. During this period, the Akkadians made a great contribution to writing by taking the Sumerian cuneiform writing and revamping it to fit their own language. It was highly useful and spread to surrounding areas, becoming a universal means of written communication. Thousands of clay tablets with cuneiform or wedge-shaped writing have been found at archaeological sites and the archaeologists can read the language. It has enhanced our understanding of the history of the region, but to underline the point about the complexity of this history, the great city Akkad, from which Sargon ruled, has never been found. By 2100 B.C.E., the Akkadian dynasty had fallen apart, possibly absorbed into the other Semitic groups, and was no more.

Q. *What civilization first used iron?*

A. Although the official "Iron Age" did not start until 1200 B.C.E., there are scattered amounts of crude iron found in many early civilizations as long ago as 4000 B.C.E., but no evidence of wide use or technology. At least some of the iron found in the years prior to 1200 B.C.E. may be an accidental by-product of copper making. Good iron was not easy to make. It required high temperatures such as 2,800°F, which was difficult to achieve. Also, the addition of carbon in the right amounts to harden the iron was not generally known. Further, quenching and reheating to reduce brittleness was only understood later. However, progress in this metallurgy around 2000 B.C.E. seems to have been accomplished by the Hittites, a diverse group of peoples living in Anatolia, a region we call today Asia Minor. The first usable iron emerged from the fire as large lumps that needed extensive beating and shaping to make a desired object. This kind of iron was called *wrought iron*. Knowledge of iron making diffused into other cultures, but the Chinese may have independently discovered its use as they were making iron bowls about 700 B.C. Among its early uses were as stirrups, bits, and other horse paraphernalia. Production of sword blades and other weaponry came later in the first two centuries of the first millennia B.C.E., with Anatolia being an important center of manufacture.

Q. *Why haven't present-day miners exploited the ancient gold and silver mines? With modern equipment they could go a lot deeper in the old veins.*

A. It is unlikely modern miners would find much more in the old mines. Going deeper is not always a guarantee of finding more ore. The reason is that many of these mines were *pegmatites* (they actually dig into the pegmatite). These are cross-cutting igneous bodies, tabular in form, connecting a molten magma toward the surface, and along which hot solutions containing the gold and silver are conveyed. These precious metals are relatively low-temperature minerals that will only precipitate when they have sufficiently cooled. In their journey along the channel toward the surface, the temperature would gradually decrease to that point where deposition could occur. Naturally, this would be the point closest to the surface. In effect, mining deeper along the vein would simply be digging along an increasing temperature gradient where the solutions would have been too hot to precipitate the gold and silver. Miners would find ample quartz and feldspar, but little of the precious metals.

Q. *Did the Trojan civilization make any significant contribution to advance science and the arts?*

A. Not much, although it inspired epic writing in the famed works of Homer, the *Iliad* and the *Odyssey*. The walled town of Troy, located in Asia Minor by the Aegean Sea, represents only one phase of a succession of at least nine or ten occupations of the site. These levels, aggregating about 50 feet in thickness, were the subject of much debate over which was the real Troy of legend. It is generally agreed that level VIIa is Homer's Troy. It rests upon a small hill called *Hissarlik* and has been dated at about 1250 B.C.E. Excavations indicate the use of the potter's wheel, and some finely worked tin, gold, and other jewelry. The Trojans also seem to have introduced the horse and horse breeding into the area. Troy was a gateway for trade goods from Anatolia into the Aegean. The ruins of its 15-foot thick walls and tower are impressive.

The fame of Troy rests principally with the well-known story of Helen and Paris. Paris, son of the Trojan king, Priam, kidnaps the beautiful Helen, wife of Menelaus, the king of Sparta, and brings her to Troy. The angered Greeks followed, under the command of Agamemnon, and lay siege to Troy for ten years. By the artifice of the Trojan Horse, the Greeks gain entrance to the city, loot, and burn it. In the late nineteenth century, the romantic and wealthy German businessman, Heinrich Schliemann, was so captivated by this legend that he undertook to find and excavate Troy. Arriving at Hissarlik, he did what no trained archaeologist would do. He cut a large swath through the center of the hill, destroying much of the archaeological evidence he hoped to find. It was an object lesson in how not to do science. He found artifacts, but there is some suspicion that he unethically "salted" the site in order to claim the reputation as the discoverer of Homer's Troy. Without Homer and Schliemann, the site of Troy today would be evaluated as a modest archaeological backwater.

Q. *The Phoenicians were a seafaring people. Who were they?*

A. They were diverse peoples of Semitic and Canaanite roots who established a series of important and autonomous cities, chief among them Sidon and Tyre, along the Mediterranean coast in what today is Lebanon and Israel. They flourished in the late Bronze Age and into the Iron Age. Throughout the Mediterranean area, they engaged widely in maritime commerce and trade, especially in cedar, ivory, and metal products. They established colonies in several places, including North

Africa. They were a talented people, and not simply skilled navigators and merchants. Their distinctive red slipware (outer decor) pottery was known to all, and fragments of it have been excavated at sites as far east as Iran. They imported copper, iron, and other metals and produced metalwork of high artistic quality. In glass blowing, they were at the forefront. Perhaps their greatest contribution to human advancement was in the development of a twenty-two-letter alphabet of such ease of use that Phoenician script spread rapidly through the Middle East and western Asia. Although not warlike, they were in conflict with the Assyrians about 700 B.C.E. and also in trouble with the Egyptians with whom they traded heavily. Still other groups like the Persians beset them. Finally, Alexander the Great conquered the region in the third century B.C.E. and the Phoenician culture was replaced by a Hellenistic culture.

Q. *Wasn't the Egyptian civilization the greatest to arise in antiquity?*

A. Certainly, the Egyptians were second to none. Even before the civilizations of Mesopotamia were just getting started, the Egyptians had already developed a calendar of twelve thirty-day months and were making contributions to astronomy. Prior to 3000 B.C.E., the Nile valley was the scene of settlement by groups trickling in from the increasingly dry western Sahara, peoples from the Middle East, and from Nubia to the south. These groups mixed with indigenous peoples and formed numerous farming communities along the Nile from the Delta, inland to perhaps the first cataract at Aswan. The fertile alluvial lands flanking the Nile, with annual renewal by flooding, assured that agriculture would be the foundation of this civilization. What is called the Old Kingdom formed when Menes, we think, united Upper and Lower Egypt. During this time, Egyptians erected enormous stone monuments with splendid art decor such as the world had never seen, including the Pyramids. Art, science, and architecture thus flourished and reached new levels. Distinctive characteristics of Egypt's culture were the focus of power in the person of the pharaoh, widespread and intricate religious beliefs, and a preoccupation with death and rebirth. These cultural traits persisted into the Middle Kingdom of the second millennium B.C.E. when Egypt demonstrated its military might and expansionist policy, penetrating and occupying areas of the Middle East. Egypt was at its height of power under Rameses II, about 1500 B.C.E. when the purported Exodus of Moses occurred. The New Kingdom brought decline of Egypt as a major power, its harassment and occupation by its enemies, and eventually its reduction to a province of Rome.

Q. *Isn't the Great Pyramid in Egypt the largest man-made structure in the world?*

A. Despite its magnificence, with more than two million limestone blocks used in its construction, it is not. That prize would have to go to the Great Wall of China. It is 1,700 miles long and would stretch from New York City to Topeka, Kansas. There is enough stone in the Great Wall to build thirty Great Pyramids. Construction began in the second century C.E. and continued over the next several centuries. The wall was intended to keep out invaders from the north, but failed in this purpose when it was breached by the founders of the Manchu dynasty. Remarkable is the fact that the wall was built up and over steep mountains, so that in places it is like climbing a stepladder to walk the wall. It is considered the longest graveyard in the world because slaves used for its construction were buried in or near the wall when they died.

Q. *Does the Great Pyramid contain mysterious mathematical truths?*

A. Built more than 4,000 years ago by Cheops, a Fourth Dynasty king of Egypt, this enormous funerary building has been used as the basis for many popular legends and claims, including its power to predict the future, preserve food, cure disease, and so on. One theory is that measurements of the Great Pyramid reveal that ancient Egyptians had a profound mathematical knowledge of the universe. It has been said, for example, that the builders used a special unit of measure, the *pyramid inch*, and that the height of the pyramid in inches, multiplied by one billion, would equal the distance between Earth and the Sun. And by calculating the volume of the pyramid in cubic pyramid inches, the result would equal the total number of all people that had lived on Earth since creation. Although such notions do stimulate our imagination, archaeologists have found that these arguments are often based on inaccurate measurements, and there is no evidence of any cosmic significance in the Great Pyramid. On the other hand, Cheops's pyramid is recognized as an amazing engineering and organizational accomplishment.

Q. *Have scientists employed instruments in an attempt to find yet-undiscovered hidden chambers in the pyramids?*

A. Yes, but in those man-made mountains, the Egyptian pyramids, it's not easy even with modern sensing devices. Some archaeologists have suspected a second hidden chamber in the pyramid of Chephren at Giza. Chephren is thought to have been the son of Cheops and the two largest pyramids in the great complex near Cairo are those of father and son. American physicists developed an instrument to send radio emissions or pulses through rock and to record any changes in density, such as would be caused by the hollow space of a tomb. No chamber was found, but the remote-sensing technique may someday aid archaeologists at other sites.

Q. *Can it be that ancient batteries found near Baghdad prove that advanced technology of aliens from outer space was available to primitive peoples?*

A. It is true that primitive batteries capable of generating one-half volt of electricity have been found, not at just one, but at more than a dozen archaeological sites. These would be useful in electroplating silver onto copper. Such devices are between 2,200 and 1,800 years old. We are faced with the following choices: (a) people back then were intelligent enough to construct such a battery, or (b) an advanced civilization reached Earth in spaceships and showed us how to make a ridiculously crude battery, and then departed. We like to think that humans 2,000 or even 5,000 years ago were as smart as we are and could make a battery. After all, it is not a sophisticated battery and would hardly be a type of battery space aliens would bring with them. When one delves into ancient technology, it is remarkable how skilled early peoples were in mechanics, art, transport, and science.

Q. *What exactly does the Egyptian Sphinx represent?*

A. The Great Sphinx represents the Egyptian pharaoh Chephren, who ruled during the Fourth Dynasty, about 4,500 years ago. It has the body of a lion and a human head, probably showing the strength and power of the ruler. It was defaced by Napoleon's soldiers, who blew off the nose with a cannon during target practice. A sphinx can be seen in the art forms of many peoples, particularly in the Middle East. The Greek sphinx was a woman with the winged body of a lion. Sometimes the sphinx also has wings. Both the Great Pyramid and the Great Sphinx have come to symbolize the great stone constructions of Egypt, and indeed, Egypt itself.

Q. *How did the Egyptians conceive the idea for mummifying their dead?*

A. The Egyptians, who were probably the earliest to do so, carried the art of embalming bodies to a high degree during the Dynastic period. Although the ancient embalmers learned how to effectively use resins and salt-containing crystals, the heat and dryness of the North African climate itself acted as a good preservative. This is indicated by the bodies of shallow desert graves that were mummified by natural desiccation, and this probably influenced the practice of mummification in Egypt. Some of the techniques were actually counterproductive, as in the case of using asphaltic oil as a preservative. This was applied to the corpse of King Tutankhamen and it aided in its deterioration.

The practice was based on the belief that the deceased's identity and spirit would endure if his body and features were preserved. Quite often the features of the person are well preserved when viewed unwrapped, and include the nose and ears. Unwrapping of a mummy is a painstaking business because the oil used impregnated the wrappings and over the centuries was *lithified*, or turned to stone. Yet study of these mummies provides useful information about the health history and nutrition of the individual and the cause of death. Many suffered from arthritis. Dental problems were widespread among the Egyptians because so much sand got in their food and wore their teeth down by subtle abrasion. Some mummies can be studied without unwrapping by simply using X rays.

Q. *Isn't there a wall in Great Britain that is like the Great Wall of China?*

A. Yes. It is neither as long nor as elaborate as the Great Wall, but is nonetheless impressive. This wall is called Hadrian's Wall and is 73 miles long, extending from the Tyne River to the Solway Firth. It was supposed to guard the northern frontiers of Rome's British province from hostile tribes to the north in much the same way as the Great Wall functioned. It contained turrets or forts at two-mile intervals along its length with a protective flat-bottomed ditch. Like the Great Wall, it failed in its purpose, being breached by attackers and destroyed twice in the second and third centuries. Hadrian, the Roman emperor who ordered the construction of the wall, was an interesting man. In his twenty years of rule, he spent twelve years away from Rome traveling through his provinces. He was a soldier-scholar of great intelligence, but spent the last several years of his reign, despite his power, a sad and lonely man.

Q. *Is there evidence of the Vikings being in North America hundreds of years before Columbus? How would the Vikings have navigated at that time since they didn't have magnetic compasses?*

A. No one can say when or where the mariner's compass was invented, but the Vikings seemed to have managed without it, since they seem to have already crossed the North Atlantic before the compass was known in Scandinavia. Even earlier, other maritime civilizations such as the Minoans and Phoenicians made ocean voyages out of the sight of land without a compass. Early seamen probably knew how to navigate using stars, sunrise and sunset, and prevailing wind patterns. However, the Vikings may have had another navigation aid. Scandinavian sagas refer to *sun stones*. Some scientists believe that these were minerals with polarizing properties. These crystalline stones turn color when pointed in the direction of the Sun, even when the Sun is obscured by clouds, as it often was where Vikings sailed. Eric the Red and his followers built two permanent settlements in southern Greenland, and their struggle to survive under harsh conditions is a saga in itself. Temporary settlements of the Vikings in Newfoundland are well documented.

13

BELIEF AND BEHAVIOR

Q. *Why do so many people seem to need to believe in religion?*

A. A belief fills a gap in our knowledge. The human mind seeks understanding of the world and a person's place in it. This is particularly so in areas presently beyond the grasp of science such as our destiny after death; whether a god awaits and judges our life and then rewards or punishes based upon what we did that was good and evil. If answers to these questions can be supplied, there is a satisfaction and a sense of security. Whether a belief is correct or not is secondary because any elements of doubt can be swept away by the invocation of faith. Often a belief can be challenged on the basis of fact and logic. In such cases, many people with strong religious convictions will deny evidence and claim that they have faith that their belief is true, regardless of the evidence.

Q. *Isn't the need to believe in an afterlife and immortality prompted by a fear of death?*

A. One of the appeals of many religions is the promise of a spiritual survival after earthly death. Certainly, few wish to die, or look forward to that moment. Like all living creatures, humans obey the dictate of nature to avoid death, survive, and reproduce. This is why throughout our history we have sought better ways to obtain food, shelter, and protection from enemies. But in the end, the face of death has a grim finality. The offer of a way out, another existence beyond death and for all time, would be the ultimate expression of survival. Hence, we have the growth and spread of religious ideas espousing this view, and the entrainment of legions of believers.

Q. *Did most ancient peoples believe in one supreme god?*

A. No. Early peoples believed in many gods, each of which had control over a separate aspect of the environment. The idea of a single all-powerful god (*monotheism*) is actually the most modern of all religious ideas, championed by Islam, Judaism, and Christianity. The belief in many gods—*polytheism*—seems to have taken root in ancestor worship, where departed loved ones became protectors watching over a family or tribe from a spiritual plane. Other gods personified different parts of the natural environment—such as wind gods, river gods, earthquake gods, and others. The first person to express an official position for a single all-powerful god seems to have been the Egyptian pharaoh Akhenaton, who perceived the sun as the personification of that god. There is suspicion that he was later murdered for this belief. Certainly not too long afterward the Israelites proclaimed one god for the Jews, but the idea that there was one universal god for all people seems to have become entrenched during the time of the early Greeks.

Q. *How did the idea of biblical creation start and become so widespread?*

A. In the Christian tradition, it is believed by most scholars that Moses wrote the first five books of the Old Testament, which perhaps were later modified in the fifth century B.C.E. by what was called the Priestly School. These books include Genesis. Many non-Christian cultures such as the Babylonians, Egyptians, and Indians have their own creation stories. These mostly predate the Christian story of creation by as much as 1,000 years. For example, the Flood of Noah is only a later version of the flood of Gilgamish from Babylonian sources. The stories are so close that in both accounts a dove is sent out to find land. Many scholars have noted the strong influence earlier accounts had on the Christian version. It is not surprising that early peoples made some attempt to explain how Earth was made and how humans came on the scene. Such concern over creation seems to have extended back into Paleolithic times. Following the time of Christ, Christianity spread rapidly into many countries and then to the New World. Naturally, Bible stories accompanied the spread of Christianity.

Q. *Weren't there geologists who claimed that Earth was once entirely covered by water? This would be proof of Noah's Flood.*

A. This seems to come closest to the theories of Abraham Werner who was indeed a famous geologist and one with many followers. A German, Werner had great charm and a magnetic personality. He attracted students from all over Europe to hear his dynamic lectures. At age thirty-five, Werner announced his theory that all of the rocks of Earth's crust were precipitated from a universal ocean. Werner did not have Noah's Flood in mind, as the event he proposed would have occurred long before any life at all was present on Earth. His idea was met with controversy, as it could not explain such things as lava from volcanoes cooling to form rock. His followers were called *Neptunists*, and ultimately his theory was rejected. However, Werner served as a catalyst for others to desert their armchairs and go into the field to establish evidence pertaining to the origin of Earth. Werner left little in the way of writing as he hated to write. In fact, he developed the habit of not opening letters because he then felt obligated to answer them. At his death, he left a small mountain of unopened mail.

Q. *Is there any place on Earth that can be historically certified as the original Garden of Eden?*

A. The original location, if there was one, is difficult to pin down. In Genesis 2, it states that the garden was "eastward in Eden." Eastward of what? Further on in Genesis, the location of Eden is associated with rivers, one of which is the Euphrates. This is a bit more real, and some scholars claim that Eden was located in the fertile crescent of the Tigris and Euphrates Rivers. For those of scientific persuasion, the hunt for the first men by anthropologists has led to Africa and certain locations such as Olduvai Gorge where humanoid bones are associated with primitive tools. The actual location of the legendary Garden of Eden will probably never be known for certain, but one can make a choice between the Euphrates River area and the sites found in Africa. Another alternative is that the Garden was a myth and never really existed.

Q. Are the Chinese the only ones who practice ancestor worship?

A. By no means. It is still an important aspect of many different cultures around the world, especially in modern primitive societies in Africa, parts of the Pacific, India, and Japan. We see evidence as far back as the Neanderthals, and perhaps earlier, that there was belief in survival after death. Ancient graves often were well equipped with food, weapons, and other possessions to aid the dead in their second life. In present-day societies, food is regularly placed at the shrines of ancestors. The ancestors, although usually benevolent toward the descendants, may at times punish them or become angry for some reason. Thus, besides affection and respect for the dead ancestor, there is often fear. In a few societies, human sacrifices have been made to the revered ancestor. Even in a modern society such as ours, the deceased are revered and idealized in a way that approaches worship. In a sense, Christian saints represent a form of ancestor worship. As persons of superior holiness in life, saints take on even greater esteem in death as they are viewed close to God.

Q. What is the meaning of the totem poles carved by Indians in the Northwest?

A. Surprisingly, this is a complicated question. The use of animal or plant symbolism in tribes or groups has been a universal practice. Through the ages, humans have had a close relationship with the animal world. They admired many of the physical traits of the animals and wished for the "agility of a cat" or the "keen eye of an eagle." Such traits today may not seem important, but in the past they were equated with survival. Some Indians believed that their ancestors could change their shape into that of an animal by magic and this took the form of ancestor worship, as a totem, within specific groups. In many cases, the totem functions as a sign of group identity, a religious symbol, and a form of protection. The word *totem*, derived from the Chippewa language, originally was *ototeman*, meaning a "close relationship."

Q. *I know Buddhism is one of the world's great religions but who was this person, Buddha?*

A. He was a real person, born in 563 B.C.E. in what is now Nepal. He grew up as a prince and son of a chieftain. Yet despite the wealth he was surrounded with, he rejected worldly goods and became a wandering monk, soon attracting numerous followers. He preached that a person passed through reincarnations of birth, sickness, suffering, and death until, through meditation and other discipline, Nirvana was achieved. Nirvana was the "supreme peace" or the parallel with heaven. Yet Buddha never insisted that the soul was immortal and Buddhism leaves considerable room for variation in beliefs. Gautama Buddha lived to be eighty years old, and his religion was well established before Christ was born. There are more monuments to Buddha than any other religious figure.

Q. *Do the Stone Age aborigines in Australia believe in life after death?*

A. They have a very elaborate system of beliefs, including reincarnation and a form of Supreme Being in the sky. We do not know what religious beliefs they held when they first arrived in Australia 20,000 to 30,000 years ago, perhaps from the area of Southeast Asia, but the environment of Australia came to shape their beliefs. Water is scarce in that country and thus a watering place is essential for survival. The aborigines believe each water hole is a dwelling place for spirits of their ancestors who are waiting to be reincarnated. Their religion also identifies closely with animals and even inanimate objects such as rocks and boulders. Their rituals include painful rites to show manhood, such as scarring of the body and occasionally knocking out a tooth with a hammer.

Q. *Was there really a death curse on all who desecrated the tombs of the Egyptian pharaohs?*

A. This idea took hold when King Tutankhamen's tomb was discovered on November 3, 1922, by archaeologist Howard Carter. Lord Carnarvon, who financed the expedition, died within two months of the opening of the tomb as a result of infection of a mosquito bite. This was followed by the deaths of the two men who photographed and X-rayed the mummy. The British archaeologist H. E. Evelyn-White committed suicide in 1924, leaving the cryptic note, "I knew there was a curse on me," adding strength to the legend. However, the chief despoiler, Howard Carter, lived to age sixty-seven, and died in 1939. Other archaeologists such as Flinders Petrie and Percy Edward Newberry, who both worked in the tomb, even eating and sleeping there, lived to ages eighty-nine and eighty, respectively. While the deaths of some members of the expedition remain a mystery, the belief in a curse is without solid proof.

No invocation of a curse was found anywhere in the tomb, contrary to some published accounts.

Q. *Why did the Egyptians mummify animals?*

A. Egyptians believed in a kingdom of the dead. Since we know that the Egyptians wanted to prepare themselves for life after death by mummification, we might assume that they simply wanted their pets along. The fact is that animals were mummified because they were considered sacred. The god Horus, for example, is portrayed with the head of a falcon, and other animals such as the cat and jackal were likewise sacred to other gods. Dogs, bulls, owls, fish, and even reptiles were mummified and given royal burials, sometimes in coffins shaped to fit them. This practice was carried so far as to mummify mice for the entombed cats to eat.

The word *mummy* originated, it seems, from a Persian word, *Amumis*, meaning "wax" or "bitumen." Centuries ago, it was believed that waxes, bitumen, and some oils had medicinal value and these were applied in an attempt to cure some ailments. In ancient Egypt, the people believed that immortality could be attained only if the body was preserved. In the preparation of the deceased for burial, bitumen was applied liberally, and even packed into body cavities after certain organs were removed. The line of thought, we suppose, was that if it was good enough to preserve the living, it was good enough to preserve the dead. The word *mummy* thus came to indicate a corpse preserved in this fashion.

Q. *Did other peoples, like the Egyptians, preserve their dead as mummies?*

A. Yes. There were mummies in many countries but perhaps not as old as the Egyptian mummies, which date back 6,000 years. In the Canary Islands, the dead were preserved in almost exactly the way used by the Egyptians of the Twenty-First Dynasty. This included a flank incision to remove internal organs for separate preservation in jars, and tightly bound bandages around the body. The Incas used similar techniques; their mummies can be found in Ecuador, Peru, Venezuela, and Bolivia. Sometimes they painted the mummies with red ocher. Mummification practices also existed in Australia. From the Middle Ages to the eighteenth century, it was widely believed that mummies had great value as a medicine and were imported into Europe for sale. Because mummies were rather rare, fake mummies, actually the bodies of executed criminals, were doctored to look like mummies and sold. Basically, in all cultures that practiced mummification, it was believed that the person would enjoy another life if the original body were preserved.

Q. *What was the Colossus of Rhodes?*

A. Rhodes is a Greek island in the Aegean Sea. It was there that, in the fourth century B.C.E., the sculptor Chares constructed a gigantic statue of the god Helios. The statue was made of bronze reinforced by iron and stood 105 feet high. Some drawings show the statue straddling the harbor entrance with ships sailing between its legs. This is preposterous. The statue was located to one side of the harbor entrance. It stood for fifty-six years before an earthquake snapped it off at the knees. The remains lay there for 400 years before the bronze was carted off on the backs of 900 camels. It must have been a magnificent sight, and is classified as one of the Seven Wonders of the ancient world. Of the Seven Wonders, only the pyramids of Egypt still survive.

Q. *Is it true that the cross, far from being a Christian symbol, was a pagan device long before Christianity?*

A. Crosses have been found painted on pebbles at archaeological sites more than 12,000 years old, and thus would have had nothing to do with the Crucifixion. Scientists believe that the cross is one of those universal symbols signifying the four main directions of the compass, and so represents all of life and everything. We know that the early Egyptians used a similar symbol with a loop at the top to symbolize life (the ankh). It is thus entirely consistent that the cross also be a symbol of Christianity. It should not be thought that a pagan origin implies barbarism or degradation. By definition, a *pagan* is a person who does not share the beliefs of Christians, Jews, or Moslems. Early on, Romans and Greeks were considered pagans because they believed in more than one god. While we might argue today about that illogic, the Greeks and Romans, as pagans, contributed significantly to science, art, and the advancement of civilization as we know it today.

Q. *Is it true that the Jews committed suicide by jumping off cliffs rather than surrender to the Romans?*

A. Yes. Masada was a strong fortress built on a steep hill in about 100 B.C.E., although wandering groups before that time may have used it. It is located along the coast of the Dead Sea in Israel. It was refurbished by Herod in 35 B.C.E., and remained a Roman fort until 66 C.E. when it was seized by Jewish zealots who were opposed to the rule and polytheistic religious beliefs of the Romans. The revolt lasted until 72 C.E. when Masada was the last stronghold still in the hands of the Jews. The Roman tenth legion laid siege there and the 960 defenders took their own lives rather than surrender when it became clear the Romans were determined to defeat them by any means. Archaeologists have found important scrolls there in the course of their excavations, and this site is visited by many tourists.

Q. *Why did Cleopatra commit suicide by snakebite?*

A. Cleopatra, queen of Egypt, was born in 69 B.C.E., and ruled during the period of the Ptolemaic or Macedonian Greek occupation. When the Romans came, both Julius Caesar and Mark Antony found her attractive and paid court. She became the wife of Mark Antony. In the civil wars that wracked Rome at that time, Octavian had managed to defeat Antony, who thereupon received a false report that Cleopatra had committed suicide; hearing this, he took his own life. When Cleopatra learned of this, she too committed suicide. She chose to die by the bite of an asp because the snake was sacred in Egypt's religion. It is of interest that Cleopatra was not Egyptian, but Greek, with some Iranian blood. Portraits of her exist that show she was not as beautiful as some say, but accounts indicate she had such charm, personality, and intelligence as to be regarded as a great historical personage. History records that she had constructed for herself a mausoleum in the royal cemetery at Alexandria, but her actual burial site is not known for certain.

Q. *Is there any natural explanation for the parting of the Red Sea during the Exodus of Moses and the Israelites?*

A. Several explanations have been offered other than the direct and miraculous intervention of God. Some historians claim the refugees actually crossed at a point of soft marshes and not open water, and the idea that walls of water were formed is just fantasy. However, at that time 3,500 years ago, one of the most violent volcanic eruptions of all time took place—the eruption of Thera, a volcanic island south of Crete. It must have darkened the entire Mediterranean and sent giant sea waves (tsunami) through the area, and this might have achieved the same results: the temporary parting of the Red Sea. It would also account for the various plagues of Egypt as recounted in the Bible because such a cataclysm would cause animal migrations en masse and other errant behavior.

Q. *How could Methuselah and others in the Bible have lived so long?*

A. Chapter 5 in Genesis mentions Methuselah's age as 969 years while Adam lived to be 930. Many other persons in the Bible had comparable longevities. It is difficult for science to reconcile such claims with what is known about human longevity today, or even in the human fossil record extending back many thousands of years. One answer might be that during biblical times, time was measured differently. For example, if a month were equated to a year, then Adam's life span would be seventy-seven years. This is reasonable. The idea that Earth traveled faster around the Sun in those days and made for a shorter year is unsupported by science. This is one of those Bible oddities for which we may never be able to find a reason or an answer.

Q. *Does anyone really know how big the biblical Goliath was, who was slain by David, or whether he even lived?*

A. The Bible claims that Goliath did exist and was about nine feet tall. There is no evidence to contradict his existence. There were, and are, human giants. In this century, Robert Wadlow would have been a modern Goliath, approaching very close to the height of nine feet. Wadlow traveled with the Ringling Brothers circus. Unfortunately, these persons suffered from hormonal imbalance that generated unusual growth. Most of these giants died very young, often at about twenty years, and despite their size, possessed no great physical strength at all. Thus, Goliath would have been a rare exception as he wore a heavy coat of chain mail, and was presumed to be a fierce combatant. Is it possible that Goliath was as weak as other giants we know about, and that the Philistines were using him to bluff their enemies? We don't know.

Belief and Behavior

Q. *Are searches still being made for the Ark of the Covenant, like in a movie a while back?*

A. Commercial films offer exciting but often-unrealistic versions of archaeological excavation. In this case the object of the hunt, the ark, refers to a chest of wood covered in gold containing tablets of the Ten Commandments. These tablets, according to the book of Deuteronomy, marked the covenant between God and Israel. By tradition, the ark was the most sacred religious symbol of the Israelites. It was sometimes carried in battle or kept in sanctuaries such as the innermost chamber of King Solomon's temple in Jerusalem. We do not know when or why this original ark disappeared. Ancient synagogues contemporary with the Roman and Byzantine empires had versions of this ark, but these were used to store the holy scrolls of the Jewish people including the Law of the Torah, when not in use, but not the tablets of the Ten Commandments. The niches in which these "arks" resided are often found in the excavations of the ancient synagogue buildings, but not the boxes themselves. There are no death curses, snakes, or other hazards (created by moviemakers) associated with these excavations.

Q. *How is it that the Romans persecuted the early Christians and now Rome is the center of Catholicism?*

A. In 312 C.E., Constantine was struggling to gain control of the Roman Empire. The Christians were a small and insignificant minority suffering periodic persecution. Constantine had seized Gaul and invaded Italy to attack his chief rival, Maxentius, who had a large army. On the eve of battle, Constantine had his soldiers paint a cross on their shields. The forces of Maxentius were defeated and Maxentius was killed. Thus, Constantine became master of the empire and proclaimed Christianity as the official religion. Years later, Constantine told his biographer that before that battle, he had seen a cross of light in the sky, and took this as a sign of divine support in his cause from the Christian God.

Q. *Does any part of the Cross of Christ's Crucifixion actually exist?*

A. According to legend, Helena, the mother of the Emperor Constantine, went to Jerusalem in the fourth century and directed excavations at the site of the Crucifixion. She found three crosses buried and identified the cross by means of its healing power. She supposedly also found the nails. What is troubling about this story is that the cross was buried. In those days, crucifixion was a common form of capital punishment and was usually meted out to lower classes such as slaves. We do not know why the Romans would have buried the crosses they used for execution. It does appear, if it was the cross that was found, that it was sliced into slivers and distributed to churches throughout Europe. Such a fragment is said to be housed in the basilica of Santa Croce in Rome. It is unfortunate but true that if all the wood around the world said to have come from the true cross were gathered together, there would be enough wood to build a ten-room house.

Q. *What is known concerning the authenticity of the Shroud of Turin?*

A. The Shroud is the purported original linen in which Christ's body was wrapped. The image of face and body impressed upon the cloth is supposed to be the only "photograph" of Jesus. The weave supposedly dates from the first century and pollen types found in the cloth are from Palestine. The image seems to have been produced by a chemical reaction between ointments (myrrh and aloes) and ammonia given off by the body. It is without a doubt the portrait of a crucified man about five feet ten inches tall with a beard; a crown of thorns; wounds in wrists, feet, and sides; and scourge marks on the back. The recent opinion of some who believe in the authenticity of the Shroud is that Christ, contrary to popular belief, was blond-haired and blue-eyed.

Skeptics point out that thousands were crucified in Christ's day, so there is nothing unusual about the image being that of a crucified man. Also, the Shroud was little heard of until the fourteenth century when it was placed on display for the price of admission. Despite radiocarbon dates from three independent laboratories that demonstrate an age consistent with the Middle Ages, many dismiss these dates as inaccurate and still believe in the Shroud's authenticity.

Q. *What was the grail so searched for by knights in the Middle Ages?*

A. There are many legends about the Holy Grail that extend back many centuries. The Grail, in the most popular version, was the cup that Christ drank from at the Last Supper. It could also be interpreted to be a dish. As such, it would be a priceless relic if it could be found. This accounts for the stories of the legendary knights of King Arthur's Round Table and their adventures in trying to find it. Doubtless, there were containers for food and drink that Christ used at the Last Supper. For that matter, any vessel used during Christ's lifetime and authenticated as such would be priceless. It is most probable that such objects have long been lost or destroyed. It should be noted that tales of sacred cups and dishes can be found in other cultures, predating Christianity and seemingly associated with early ideas of immortality.

Q. Does the water from Lourdes have magic curative power?

A. Yes and no. Bernadette of Lourdes was a child of fourteen in 1858 when she claimed to have received several visions of the Virgin Mary at a grotto there and found a spring that has been flowing since. Today, the town receives over two million visitors each year, many of them hoping to be cured of ailments. Yet chemical analysis of the springwater shows it is good drinkable water without any special properties. Vials of this water are sold with the idea it will cure diseases and other infirmities. Many apparent cures turned out to be temporary. The Church itself is rigorous in its judgment of claimed healings at Lourdes. And there are few such cures, considering the millions of people who have been there. On the other hand, the powerful effect on the mind and spirit—the psychology of it—may well have a beneficial effect for the afflicted. Bernadette herself lived only to age thirty-five, dying in a convent in great pain.

Q. Why was Joan of Arc burned at the stake?

A. It is a complicated question, but one with political overtones. In the early fifteenth century, with the Hundred Years' War raging, Joan was a young peasant girl thirteen years old who began to hear "voices." Later, she gained an audience with the French leader, later Charles VII, who was convinced by her divine inspiration and permitted her to lead French troops against the English, resulting in a string of victories for the French. Joan herself was a tough cookie, leading the armies with sword in hand and being wounded more than once. She was eventually captured by the English and put on trial. She was found guilty, among other things, of wearing men's clothes and also of hearing the divine voices in French rather than in English. Because she was an idol of the French and a rallying point, she was burned.

Q. *Can it be said that the Crusades of the Middle Ages to capture the Holy Land achieved their purpose?*

A. No. It might be better described as a bloodbath for both sides extending from the year 1095 to 1270 in which eight major expeditions to the Holy Land took place. It should be noted that sites such as Jerusalem were just as sacred to the Muslims as they were to the Christians and the Jews. The first crusade of united European Christians seized much of Palestine including Jerusalem, where they besmirched themselves by putting both Jewish and Muslim inhabitants to the sword. After about forty years of Christian occupancy, the Muslims counterattacked to reclaim territory and a second crusade was sent at the urging of Pope Eugenius III. This ended in a humiliating defeat of the Christian armies and the pope called for a third crusade under the leadership of Richard the Lion-Hearted. This ended in a stalemate and a truce with Saladin, the talented leader of the combined Muslim forces. A fourth crusade in 1198 was undertaken to seize Egypt but instead the army turned against the Christian city of Constantinople and sacked it. The Children's Crusade fared no better, as thousands of children were either lost or sold into slavery. This crusade helped to initiate the fifth crusade. Subsequent crusades all ended in failure. The crusades were not a very noble chapter in our history. Ironically perhaps, the most "Christian" figure was Saladin, whose troops reoccupied Jerusalem in 1187 without bloodshed or looting.

Q. *How did the custom of having wedding cake at a marriage originate?*

A. It probably got started in a cornfield. In the early stages of man's agricultural developments, a bride would be adorned with ears of corn as a symbol of fertility. Later, this pagan custom changed somewhat when they started using small pieces of cake which were not primarily for eating but were broken up and tossed over the heads of the couple. In some cultures the cake was worn on the person. Later, cakes were made, sometimes with both sweet and bitter ingredients to symbolize the good and the bad that a married couple would encounter. This cake was eaten and evolved into the modern wedding cake which is, optimistically, all sweet. Even so, the wedding cake is still a symbol of fertility. The honeymoon is a modern custom. In former times, family and friends followed the newlyweds right to the bridal chamber. No wonder somebody finally thought it up.

Q. *I've noticed that most graveyards are on hills. Is this because it is closer to heaven?*

A. Various symbolisms have always been seen in the landscape, which is highly diverse. Hills, especially impressive mountains, represent a higher spirituality and something difficult to reach as opposed to valleys which are earthly and usually more comfortable. The old Greek gods lived on Mount Olympus and Moses received the Ten Commandments on Mt. Sinai, yet some mountains had an unsavory reputation. Even the wicked witch in *The Wizard of Oz* perched her castle on a mountain. Man has also made artificial mountains such as the pyramids of Egypt and Central America, both associated with life and death. Although not all cemeteries are located on hilltops, as you observed, many are. In many graveyards the deceased are buried in an east-west position so that if they awaken, the first thing they would see is the rising sun. The beliefs of many may be a way of helping the deceased with the first few steps toward a better life beyond.

Q. *Why does the new bride get carried over the threshold?*

A. It goes back to the Bronze Age, especially among the early Greeks when each clan had its own gods for protection. Marriage was forbidden within the clan. So if you got married, it would be with someone from another clan. Of course, this meant that a woman would shift over and become part of a different clan after the ceremony. This was fine, but there was a possible obstacle. The woman's clan gods might get angry that she was leaving them in favor of another clan and other gods. Therefore, a fake battle was staged to suggest that the woman was not leaving her clan of her own free will. The battle ended with the groom carrying the woman over the threshold of the new home as though she were protesting the whole thing. The gods were thus pleased, and harmony was assured. At least for a while.

Q. *When did man first start drinking coffee and tea?*

A. Their origins are rather vague. Both coffee and tea are stimulants owing to the caffeine content. One legend has it that goats were observed munching on coffee berries and then frolicking in a manner rather unlike goats. So the goatherd tried it. This was about 1,200 years ago in Kaffa, Ethiopia, hence the name. Later, it was learned how to steep coffee into a brew. Tea apparently is a much earlier beverage. Existing records indicate that it was being cultivated as early as 350 B.C.E. Its origin is northern China, from which it spread to Japan in the sixth century. The East Indian Company with its clipper ships introduced the oriental brew into England and colonial America between 1600 and 1858. The tea tax that England imposed on Americans led to the famous Boston Tea Party, contributing to the American Revolution. The drinking of these beverages has become an almost religious ritual, for tea in the Orient and England, and coffee in the United States.

Q. *I have often heard the expression that someone "ought to be boiled in oil." Was this ever done to some unfortunate people?*

A. Yes, many times, and worse. A variety of unspeakable tortures have been practiced, extending back into ancient times. Use of such cruelty, usually ending in death, was justified as in the case of extracting a confession from a criminal. Later, torture was used against political enemies or those opposed to established religion as in the case of the infamous Inquisition. And torture was also employed for revenge. The most repugnant application of torture was for "sport," as in the Roman games under such emperors as Nero. A variation of the boiling in oil torture was to suspend a person over a fire and baste him with oil. On one occasion, Nero had five thousand women, men, and children thrown to wild beasts during a thirty-day period. While such cruelties are no longer widespread, torture of enemies with more sophisticated methods such as brainwashing and electrical applications still takes place in different parts of the world.

Q. *How accurate is the lie detector? How does it work?*

A. The first thing we should understand is that the so-called lie detector, or polygraph, does not detect lies. It records human emotional responses in terms of changes in heart rate, perspiration, muscle tension, and so on. These responses must be interpreted by an operator skilled in the use of the polygraph, and not everyone is. Let us suppose a person sincerely believes he has seen a flying saucer and is given a lie detector test. Although the flying saucer does not exist, the lie detector will indicate the person is telling the truth. Sensationalist writers seize upon this as "proof" that flying saucers exist. Not so. Another thing to be considered is that we all have guilt feelings about events in the past, and a question may touch upon these feelings and create an indication of guilt that can confound even an expert. While the lie detector may have its uses, it is by no means a magical all-knowing instrument.

Q. *Where did Grimm get his ideas for the fairy tales children still read?*

A. Actually, there were two Grimms—brothers named Jacob and Wilhelm—who wrote the fairy tales more than one hundred years ago. *Grimm's Fairy Tales* were a by-product of their professional interest in language, grammar, and history. From the time when both were university students, they were fascinated by folklore and mythologies that had been handed down by word of mouth in Germany. They avidly collected these tales from whatever source and fashioned them into readable stories. The stories encompass the magical lore of many lands, including elves, giants, and animals transformed into people and vice versa. The Grimm brothers would have been famous even if they had not written the fairy tales. In the early nineteenth century they undertook to compile a great dictionary of the German language, a task so difficult that it was not completed until 1960 by other scholars. Both brothers worked closely together all their lives and are buried side by side in Berlin.

PART 5

THE
PARANORMAL

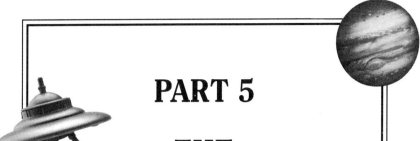

14

MAGIC, WITCHES, AND DEVILS

Q. *Do the words* paranormal *and* supernatural *have the same meaning?*

A. If we take the words *natural* and *normal* to have pretty much the same meaning, and add the prefixes *para* and *super*, then nothing is very different. These prefixes suggest something above or in addition to the norm. Yet current usage separates these terms somewhat. *Supernatural* has always implied such mysterious things as ghosts, haunted houses, and "things that go bump in the night." These sorts of happenings are generally beyond science's present capability to study them. On the other hand, the word *paranormal* has frequently been used to describe such things as mental telepathy, clairvoyance, *psychoki-*

nesis (the ability of the mind to move objects), and *precognition* (knowing things in advance). These latter categories are indeed being studied through scientific methods. Together, they constitute what people think of as *ESP* (extrasensory perception). Although the distinction is a little hazy, you might think of the paranormal as things amenable to study by science, and the supernatural as being yet outside the bounds of science.

Q. *Aren't magic and science closely related?*

A. We would probably agree, but only in a certain context. If you could (magically!) transport yourself back to the Middle Ages, and show people of that time a tape recorder or a cigarette lighter, they would doubtless call it magic. You would be proclaimed a magician or maybe a god. Or less fortunate, you would be thought of as a demon and be burned at the stake. The point is, whatever is not understood, or does not seem to conform to the laws of nature, is often viewed as some kind of "magic." It is an attempt to understand. All cultures through the ages evolved their own superstitions and magical practices when confronted by the unknown. As knowledge advances, much of what was formerly "magic" becomes scientific truth, or alternatively is discarded as false.

Q. *What exactly is magic, and does it really work?*

A. To define magic is like trying to nail a custard pie to the wall. There are many definitions, many practices, and for different purposes. The use of magic seems to be universal, being practiced in all cultures past and present. We think there are some common threads here. Most cultures employ objects or other kinds of symbolism together with incantations. These are used in various ways in an attempt either to effect some degree of protection against real or imagined evil, or to punish or harm some enemy.

Does it really work? It seems to be a matter of belief. If you believe that a witch doctor can put some kind of "hex" on you that is harmful, even causing death, it could happen. There are documented cases in Africa. On the other hand, if you scoff at something like a hex, and one is invoked, can you be harmed? Probably not.

Q. *Do fringe ideas and extraordinary claims in any way aid the scientist?*

A. A whole book could be written on this subject since the area is so vast, but in general the answer is yes. Take, for example, geographical myths. It was once widely believed that there was an undiscovered Northwest Passage to the Orient and also, in the South Pacific, a great southern continent. Neither of these exists, but the belief that they were there somewhere, yet to be discovered, influenced geographers and explorers to search and discover the true nature of the physical world.

Other esoteric claims such as Bigfoot, pyramid power, and ESP might someday advance modern science. The borderlines of scientific knowledge do change.

Q. *What does that magic word* **abracadabra** *mean?*

A. This word has been in use for nearly 2,000 years, having originated in Roman times. At that time, people believed in a god called *Abraxas*, who could help shield a person from evil if this god's name were inscribed on stones or jewelry and worn on the person. Use of the word *abracadabra* was thought to cure illnesses, especially fevers. The word was uttered often during the Middle Ages at the time of the bubonic plague in Europe to ward off this dread disease. Nowadays, stage magicians use the word in their illusions of making things vanish or appear, or in other tricks. Thus, the word has lost its original meaning. Nevertheless, the next time you get sick, say "abracadabra."

Q. Wasn't Cagliostro an infamous and evil magician?

A. Alessandro di Conte Cagliostro was probably not a bad person, but was condemned mostly because of jealousy and suspicion. As we understand it, he was a man of active intellect interested in the occult and the supernatural at a point in history when such things were frowned upon by the Church. Born Giuseppe Balsamo in 1743 in Sicily, Cagliostro grew to adulthood gaining skill in such matters as hypnosis and grasping the elements of psychology such that he effected some marvelous cures of sick people and gained wide fame. He also dabbled in alchemy, trying to change lead to gold (although he was rich and needed no money), and in other occult matters. Thus, he was held to be a magician. This was dangerous. Unjustly accused of the theft of a royal necklace in France, he went to Rome, where he fell into the hands of the Inquisition. He languished in prison until 1795 when, according to some reports, he was strangled by his jailer at the age of fifty-two. To some he was a charlatan, to others a showman. Cagliostro was not evil, but rather sought to help others. He lived at the wrong time in history.

Q. What does a witch do in the performance of witchcraft?

A. A witch performs acts to harm or injure an enemy, or acts to achieve other goals not necessarily evil. For example, a man may wish to have a specific woman love him. From Paul Christian's book *The History and Practice of Magic*, consider the following advice: "Take a dove's heart, a sparrow's liver, a swallow's womb and a hare's kidney, and after having dried them, reduce the whole to a very fine powder, to which you will add an equal quantity of your own blood and leave the mixture to dry. If you make the person you desire eat it, she will not be able to resist you for very long." There is no scientific basis to suggest that such a concoction would work, and there might also be some objections by the sparrows, doves, and other animals that become party to this plot. Other tools of the witch include spells and incantations. Note that in our example above, we see the practice of good or *white* magic wherein a witch assists a man in winning a woman's heart. Not all witches are evil. As the 1939 classic movie *The Wizard of Oz* depicts, there are good witches and bad witches.

Q. *Did most of the executions for witchcraft take place in England?*

A. By no means. It was a madness that engulfed much of Europe as the Middle Ages came to an end. Some of the most horrible stories on record come from Germany. In a thirteen-year period in the state of Bamberg alone, three hundred unfortunates were put to death, earning that region the title of the "shrine of horror." While there may have been some motivation to save the souls of these people, other motives played a role as well. Authorities confiscated the property of those executed, and woodcutters who sold wood for the burnings at the stake became enthusiastic accusers. It seemed that for a time no one was safe. If a person expressed the slightest hint that another was a witch, such persons themselves often were arrested and tortured. Under severe torture, most people will confess to anything. The Roman Catholic Church often is singled out as responsible for all these killings, but it should be noted that the Protestants also burned people at the stake for witchcraft.

Q. *How can the fact that many sick people are cured by witchcraft be explained?*

A. There are logical reasons for such recoveries. The human body is certainly very durable and often recovers using its own natural defenses, as it did for thousands of years before modern medical science. Most witch doctors are well aware of the curative powers of certain herbs and other concoctions and also the value of massage and other physical therapy. Perhaps most important is the psychological factor: the most dramatic recoveries are among those with the greatest conviction and belief that witchcraft does the work. It is in essence what is called the *placebo effect*. Thus, when a healer acts to cure a person, the fear of illness, which has often caused the illness, is dissolved and the person is on the road to recovery. Be that as it may, we still recommend that you see your physician when you are sick.

Q. *Where did the idea originate that sticking pins in a doll by a witch could harm the person the doll represents?*

A. This belief cuts across many cultures and seems rooted in childhood perceptions. To the child, the doll was real and could be scolded or praised or punished. At the adult magical level, the doll is made of wax, clay, or other material, and made to resemble the living person as closely as possible. Incorporation of blood, hair, nail clippings, or other objects related to the victim makes the magic even more powerful. Pins or thorns thrust into any part of the doll will inflict pain or destruction on the corresponding anatomy of the real person. To the anthropologist, this type of witchcraft is part of what is known as *imitative magic*. In modern garb, athletic teams choose names like Tigers or Bulls, as if to embody the strength of the suggested animal. To some, the doll serves as a focus for the hate of the witch. Indeed, in some societies where this form of magic is believed to succeed, it has been known to have worked, even causing death. It may also be used to make a man love a woman rather than inflict physical harm.

Q. *When was the last time a witch was burned at the stake?*

A. Witchcraft has always been regarded as a practice of black and thus evil magic, wherein persons enter a pact with the Devil to gain power to carry out evil purposes. During the 1600s and 1700s, there was almost a paranoia against witches not only in the United States but also in Europe. People were hunted down and tortured to obtain confessions that they were witches. Under severe torture, most "confessed." In Europe, the last official burning at the stake occurred in Scotland in 1727. In colonial America, several people were put to death in 1692 in Salem, Massachusetts. After these times, official government-sponsored witch-hunts declined. Yet as late as the 1950s, two people were hanged in Mexico who were thought to have been witches. Thus, even today, somewhere in the world people are still probably accused of being witches, and perhaps losing their lives as a result.

Q. *How did the belief in witches actually begin in Salem?*

A. In general, ideas and beliefs about witchcraft among the people of Salem were influenced by European attitudes toward Satanism. The Salem witchcraft episode seemed to start in the late seventeenth century with the arrival of a new pastor, Samuel Parris, together with his wife, daughter, and niece. They were accompanied by two young slaves, one a young West Indian girl named Tituba whose father had been a tribal witch doctor. In what might be called an early American slumber party, a group of young girls began meeting with Tituba in the evenings in order to practice "magic." History records that shortly after these meetings, the girls exhibited a pattern of strange behavior, having hysterical fits and convulsions. Doctors could find no physical cause and concluded that the children were bewitched. This belief spread and eventually over two hundred persons were thought to be possessed by evil spirits. As a result of the Salem Village witchcraft trials, nineteen people were put to death in 1692, not including two dogs that were hanged as witches.

Q. *Witches are supposed to have "familiars." Are they demons?*

A. They were thought to be demons or imps by the witchhunters. They took the form of a small animal such as a mouse, rabbit, frog, or toad, but most often it was a cat. It was believed that the Devil had given or sold the familiars to the witches, to assist them in their work of casting spells on innocent people. The idea of a witch's familiar seems to have originated in England around the fifteenth century because it is noted that in stories of witches of other lands, familiars are hardly mentioned. Perhaps certain elderly women in their loneliness became strongly attached to a pet cat or other animal. When they were accused of witchcraft, the cat got the blame too. Nowadays, it seems hardly possible that such superstitions led to the torture and execution of numerous women, and some men as well.

Q. *Can a cat suck the breath out of a small baby?*

A. No. The exhalations of any human being consist of carbon dioxide and no air-breathing animal would have need of it. It is a superstition that may have originated when, years past, a crib death occurred and a cat was nearby. Also, a cat may curl up to a person—baby or adult—simply to enjoy the warmth of the body on a cool night. Cats, especially black cats, have been victimized throughout history, and even executed for imagined evils, including association with witches and the Devil. Yet in some cultures, a black cat is considered a symbol of good luck and a house is lucky that houses a black cat. But don't let one cross in front of you!

Q. *My grandmother claims that her cousin was killed by the evil eye in Italy. What exactly is this evil eye?*

A. The evil eye is one of the most widely believed ideas in the world. In one form or another, this belief is found in much of Africa, the Mediterranean, and Europe, although it is not found much in the Pacific. The basic idea is that some people or animals can cause harm simply by glancing at other people or property. Sometimes the evil eye is cast maliciously upon an enemy, but other people can have the evil eye involuntarily like a sickness. Certain people are often avoided for fear they have the evil eye, such as one-eyed or cross-eyed persons, hunchbacks, or others with some deformity.

The origins of this belief are not known precisely. As is well known, the eyes can reveal emotions and hidden feelings, and are called the "window of the soul." We might infer that cases of anger or hatred projected from the eyes to another individual followed by some coincidental catastrophe gave rise to the belief.

Q. *If someone puts a curse on me, what can I do about it?*

A. It might be wise to do nothing. A curse seems to have an effect only if the recipient of the curse really believes it will. In some societies where such belief is universal, everyone avoids a person who has been cursed. He becomes a social leper, an outcast. Psychologically, the effect is devastating and often lethal. Yet most curses are not invoked to kill, but to punish. A man may see his cattle die one by one, and know he is under a curse. As late as the middle of the twentieth century, a man in Arizona killed a person he thought had placed a curse upon his wife. In ancient Egypt, curses were made against anyone who desecrated the grave of a pharaoh. Thus, a curse was a form of protection. In effect, a curse is only as powerful as the depth of belief in its power, and clearly has psychological roots. If you take our advice, however, try to avoid being cursed.

Q. *How did the idea originate that fairies exist?*

A. The belief in such creatures extends back at least many hundreds of years and its origin is a complex matter. One interesting theory is that during the Stone Age, new settlers in a region displaced the original inhabitants, who then retreated to isolated areas and were not seen very often. Because of their greater familiarity with that region, they were held to be elusive and the idea grew that these people possessed magical abilities. Often, the displaced people were of shorter stature than the invaders, and so the concept of fairies as "little people" became widespread. While modern fairy tales portray fairies as good little people, the earliest beliefs show them as sinister and evil, or at least mischievous. An added reason for believing in fairies could have been that when something bad happened, it could be blamed on the fairies rather than on one's own blunders.

Q. *Is it true that Arthur Conan Doyle, the creator of Sherlock Holmes, believed in fairies?*

A. It is true that some little girls who took pictures of pasteboard drawings with small winged fairies apparently duped Doyle. He concluded that they were genuine. It is ironic that Doyle could be so fooled in light of the precise logic and attention to evidence shown by his alter ego, Sherlock Holmes. Indeed, although fiction, Doyle's detective stories about Holmes helped to advance the science of criminology. Yet Doyle wrote the Sherlock Holmes stories in his twenties and it was only later in life that he became interested in spiritualism. His conviction about this is illustrated by an alleged communication from the grave from the Great Houdini's mother, which was in English. Houdini protested that his mother could not speak English. Doyle concluded that Houdini's mother had learned the language while in the spirit world. While one can be rational and logical, one can still be gullible.

Q. *How did the idea of the Devil originate?*

A. The idea of the Devil today is associated with early Christianity as there are several references to the Devil in the Bible. However, the idea may go back much further to the time of cave-dwelling primitive man. Cave drawings thousands of years old depict a man wearing animal skins and a pair of horns. These drawings may be the forerunner of the Horned God that represents the forces of nature and is the earliest known deity. This god symbolized both good and evil. As civilizations developed, the notions of good and evil became more crystallized, leading to separate personifications for good and evil. The idea today that the Devil has horns, hooves, and a tail may be traced back to the Horned God. The Devil has been called by many names because almost all societies and cultures symbolized evil in some way.

Q. *Why is the Devil depicted as having a pair of horns?*

A. From primitive times when man, the hunter, stalked animal herds, the bulls and rams—males that possessed horns—were seen to be the biggest and most powerful. Thus, horns became symbols of courage and strength. They were also symbols of fertility as they battled for the females of the herd. Early men wore animal skins and horns in hoping to strike fear into their enemies.

As pagan religions developed, the gods had horns as beings both feared and admired. As one religion came to replace the older one, the previously worshiped gods became the demons and evil ones of the new. The casting out of the Devil from heaven can be interpreted as a former god, the angel Lucifer, evolving into the symbol of evil of the new Christian religion. It was in the fifth century that the Church declared that the Devil had horns.

Horns are made up of hardened protein and are a permanent feature of the animal. In contrast, the antlers of a deer are not true horns because they are made of bone and are shed annually. The vernacular term "horny" suggests an aggressive male at large, seeking female companionship.

Q. *Why is the goat so often identified with the Devil?*

A. The origin of this idea is obscure. Goats were worshiped in Egypt more than 2,500 years ago, but not as an evil entity. During the many witch trials of the Middle Ages, accused witches confessed to confronting the Devil taking the form of a goat, and hearing the goat speak like a man. There is some suspicion that sometimes the goat was actually a man clothed in animal skins. Adding to the idea of the Devil as a goat is its reputation as a rather dark, smelly, and repugnant animal. This is an undeserved perception of the goat, which through many centuries has been a very useful animal for humankind. Early on, the skins were used for leather and clothing, and the milk and meat for food. And we still use the domesticated goat. Goat's milk is very good and more digestible for infants than cow's milk, and its cheese, feta, is the glory of a Greek salad. Some goats can produce more than 6,000 pounds of milk in a year and the meat has a delicate flavor similar to lamb. Yet the goat has had a devil of a time gaining any general admiration.

Q. *Why does the Devil carry a pitchfork?*

A. The exact origin is unknown. We might surmise that it started out as a male symbol in the same way that the broomstick was the symbol of a woman; that is, in olden times the tool used to carry out domestic chores. The pitchfork represented the man, working in the fields. Somewhere along the line, these two symbols became associated with witches and demons, perhaps because they were carried to nightly gatherings of alleged witches and warlocks. A more obvious use for a pitchfork by the Devil in the realm of hell would be to prod and torment the condemned souls there.

Q. *Why are cats thought to have nine lives?*

A. As one of the earliest carnivores domesticated by humans, the cat has been observed for many centuries. By nature a solitary, nocturnal hunter, the cat comes and goes often, sometimes for days or weeks. After a long absence, a returning cat may have been regarded as returning from the dead, leading to the idea that it had more than one life.

Although the cat was worshiped as a god in ancient Egypt, it has more often been associated with the Devil, especially black cats. Cats were frequently burned alive or boiled in oil by witch-hunters. As to why a cat should have nine lives, consider that in the Bible the "number of the beast" was 666. The superstitious would say, 6 + 6 + 6 = 18 and 1 + 8 = 9. That is why the Devil's agent, the cat, has nine lives. Also, a witch can transform herself into a cat only nine times. You can do anything you want with numbers, but it does not necessarily lead to truth. We like cats.

Q. *Why are certain musical instruments, such as the flute or violin, associated with the Devil?*

A. Music in various forms and instrumentation has been associated with nearly all religions for thousands of years. Music, and in some cases dancing, has the power to evoke emotion and ecstasy, bringing the participant into a perceived spiritual state. This is all well and good, but as various religions developed, some sects were thought to be more in league with the Devil, and their ritual music regarded with suspicion. Thus, the violin and flute (associated with the pagan god Pan) were viewed adversely. Indeed, at one time, quite an argument took place over the use of the organ in Christian ceremonies. Drums are another instrument generally associated with non-Christian ritual. Before the Christian era, prophets used drums, singing, and dancing to induce the physical and mental state that enabled them to see into the future. The state induced was more likely one of self-hypnosis.

Q. *Were the reported cattle mutilations caused by UFO aliens or witches' cults?*

A. Published reports of cattle and sometimes other farm animals being found, particularly in our western states, with tongue, eyes, and sex organs removed have appeared at intervals over many years. To lay the blame on aliens is a rather foolish hypothesis. Yes, there are cults of various kinds, proclaiming allegiance to the Devil, which use animals or animal parts in some of their rites. It would seem far easier for the cultists to pick up a stray alley cat for their rituals rather than chase some rancher's cattle out in the middle of nowhere on a dark night. What sensationalist writers fail to mention in their stories on mutilation is that the animals involved have usually been dead from normal causes for several days before the owner discovers them. In the meantime, smaller animal scavengers have probably attacked the carcass, eating the softest parts such as the tongue. This is another case where something fantastic is proposed, or something sinister implied, when there is a perfectly natural explanation.

Q. *What is the explanation for the mysterious footprints made in the snow one night in Great Britain that were seen for many miles and thought to be the footprints of the Devil?*

A. There were many in Devon who woke up that morning in February 1855 and believed the Devil had strode through their towns and villages. There were prints in the snow of cloven hooves similar to those of a donkey, perhaps three or four inches in diameter. These prints were spaced about eight inches apart and incredibly trended in a straight line through streets and backyards, over fences, and even crossing snow-covered roofs of houses. The people of the region assumed at first it was some kind of animal or perhaps several that had romped through their area during the night. But what donkey or other creature could move over eight-foot walls without breaking stride? These unusual tracks extended over 100 miles and even crossed two miles of open water in an estuary. Our best guess is that the county of Devon was visited that night by a meteor shower, with small fragments that impacted silently and made impact markings that, while in places resembled prints of a hoofed animal, also resembled some impact craters on the moon. At that time, most people did not believe the assertions of some scientists that stones could fall from the sky. An athletic devil seemed more plausible.

Q. Why is the jackal often regarded as a creature of the Devil?

A. Probably because it is a hunter by night, an eater of carrion, has a reputation for cowardice, and has a disquieting wail. Also, the jackal has long been associated with death going back to the time of ancient Egypt. The Egyptians personified the jackal as Anubis, the god that guarded cemeteries and guided the deceased to judgment.

There is no basis for considering the jackal a coward. In packs, they can chase down and subdue much larger animals such as antelope and sheep. To eat the leftovers of a lion's feast is an easy meal. It makes good sense to the jackal. After all, who wants to argue with a large formidable creature like a lion when you (the jackal) are only two feet long (neglecting a foot-long bushy tail) and weigh less than 25 pounds? The jackal is much more doglike than the hyena, with which it shares an undeserved unsavory reputation, and indeed, the jackal can interbreed with domestic dogs.

Q. Why do some people cross their fingers when they are hoping for something?

A. This might be traced to early Christian times when people made a cross with the forefinger on the forehead to provide protection and interference from the Devil. Crossing the fingers nowadays is still a form of invoking protection. There are any number of superstitions associated with the fingers. It is considered rude to point with your finger, and this may stem from the idea that something bad will happen to whatever is pointed at. In some primitive societies, as a sign of mourning, a finger was amputated after someone died. There are also separate beliefs concerning the fingernails. They should never be cut on Sunday or the Devil will have you all week; parings should not fall into the hands of a witch, who can make use of them in casting evil spells.

Q. *Was there a real Faust who made a pact with the Devil?*

A. Faust was a real person who lived during the sixteenth century, dying about 1540. What is confusing is that some reports were of a Johann Faust and others of a Georgius Faust. There are only so many possible explanations to this discrepancy. It may have been two different people entirely. They may have been twins or the same person with a dual personality. One was a doctor of theology, while the other apparently had his hand in black magic. Perhaps he purposely changed his name to deceive others, which would fit his scheming personality. Both have been reported as charlatans.

According to most accounts, Faust was a good-for-nothing braggart who claimed magical power and performed tricks for audiences around Germany. Among his claims was that he had a close relationship with the Devil, and referred to the Devil as his "crony." This was probably pure baloney to impress naive audiences. No doubt Faust would have faded fast into history but for the fact that Protestant churches seized upon Faust after his death as an example of what can happen—eternal damnation—if a person sought too much knowledge. Books and plays were written on this theme, especially by Christopher Marlowe and later Goethe. However, Goethe modified the theme somewhat by showing that even an evil person such as Faust might be forgiven by God. It is interesting that Faust, who himself contributed little to humanity, became an inspiration for a body of outstanding literature.

Q. *Are there genuine cases of demonic possession as shown in the movie* **The Exorcist?**

A. The movie illustrates very well the symptoms of possession that apparently are manifested in such reported cases. These include physical effects such as vomiting and frenzy, and mental effects such as personality change, use of filthy language, and so on. People have believed in demonic possession for many centuries. Why is "God bless you" said after someone sneezes? It was believed that the soul was expelled by a sneeze and at that moment a demon could enter and take control of the body unless a blessing was given. Most of the cases of possession we have heard about involve individuals who are deeply religious and believe strongly in demons. For this reason, some cases might be due to strong autosuggestion. Psychologists have noted that symptoms of possession are very similar to those of persons who are mentally ill or disturbed. Exorcism parallels psychiatric treatment of mental patients. Many cases of possession took place long ago and are poorly documented. These cases can't be evaluated properly.

Q. *Just what are demons? Did demons in real life create the havoc of* **The Amityville Horror?** *Are they like ghosts?*

A. The word *demon* is a general term used to describe an evil spirit, although in some cultures a demon could be a good spirit. Indeed, they have been regarded as the ghosts of humans seeking revenge for alleged wrongs, and are able to take on bodily forms such as werewolves, harpies, and vampires. Demons of nonhuman origin have been depicted as followers of the Devil, who himself is the supreme demon. Demons have the reputation of lurking at night in forests and byways to pounce upon and harass travelers. Mentally disturbed persons in the

Middle Ages (and even today) were thought to be possessed by a demon and required exorcism. If it did in fact exist, the so-called Amityville Horror would be considered a demon. Nobody we know has captured and dissected a demon, so the concept is without scientific credibility. Nonetheless, the subject of demons makes for some good, hair-raising stories to be told on dark and stormy nights.

Q. *How did Halloween originate?*

A. It was a pagan celebration among the Celts and other peoples in northern Europe well over 2,000 years ago. It was regarded as the time of the coming of winter and it was believed that the dead returned to their former homes looking for food and warmth. Thus, it came to be that food or other offerings were placed on doorsteps for the dead. Today's trick-or-treaters represent these ghosts looking for a handout. Demons and witches were also believed to roam the darkness on this night, and real witches and their covens continued to celebrate it into modern times. About 1,300 years ago, it was made a Christian feast day honoring saints and martyrs and called All Hallows' or All Saints' Day, perhaps to counter its pagan origins. The practice of children going from door to door for treats is a fairly recent custom, one that seems to be threatened by some of the offerings of mentally unsound persons.

Q. *Is Hades the same as hell?*

A. It has come to mean that in modern times but they were not always considered to be the same. *Hades* was actually a person or god in Greek mythology, otherwise known as Pluto, who ruled over the underworld, which was peopled by the dead in a realm of cold and darkness, rather than flames and torment. The word *hell* comes from an Anglo-Saxon word meaning "to conceal or to cover up," as in the case of a grave. The concept of hell prevails in many cultures, and it is generally thought of as a place of fire and torment, often eternal but not necessarily so. The earliest ideas of the hereafter were conceived in terms of heaven. Later, the idea of an infernal region of punishment for the wicked took form when it was perceived that some people regarded as evil were going unpunished in this life.

15

MONSTERS

Q. *Is the Loch Ness Monster the only monster reported in Scotland?*

A. The Loch Ness Monster is far from being the only monster reported. It seems every lake in Scotland has a resident monster. Examples include the monsters seen at Lochs Lomond and Lochfyne, and one from Loch Awe resembling a giant eel and described as "big as a horse with incredible length." If you look anywhere in the world where there is a sizable lake, there is also a monster: Ireland, Canada, Australia, Mexico, and so on. There are monsters in most lakes in the United States such as the one in Lake Walker, Nevada, that (so it is said) is a man-eater, but eats only white men by special agreement with the

Indians. Most monsters are not bad and have affectionate nicknames such as Nessie in Loch Ness and Champy in Lake Champlain, New York. Such monsters are created by mistaken identification or imagination, to increase the tourist trade, or are made up just for fun. In 1934, Sir Arthur Keither said, "I have come to the conclusion that the existence or nonexistence of the Loch Ness Monster is a problem not for zoologists but for psychologists."

Q. *If there actually is a Loch Ness Monster, why hasn't it been caught?*

A. There is enough good eyewitness testimony to conclude that there is something there, but what? A favorite theory is that it is a plesiosaur, a sea-going reptile that became extinct 70 million years ago along with the dinosaurs. Scientists don't think so. The plesiosaur lived in shallow salt water and was an air-breather. The loch is deep fresh water. Sightings of the monster are too infrequent for an air-breathing animal; there is only one sighting for every 350 hours of observation. Most sightings have been of snakelike humps breaking the surface of the water. It could be an unusually large eel. There are many eels in the loch. They are deep, dark-water dwellers and are not air-breathers. To catch the monster, if it exists, is a tall order. The water is stained dark brown by peat, so visibility is poor. Scuba divers can see only a few feet in any direction. Water depths reach 950 feet. The monster has a secure hiding place, even if an extensive effort is made to catch it.

Q. *If Bigfoot exists, why haven't any of its bones been found?*

A. Bones are a crucial point in the Bigfoot debate. Like Nessie in Scotland, the evidence for Bigfoot tends to be insubstantial and disconnected. The presumed habitat of Bigfoot is the Pacific Northwest and the West Coast of the United States, an area visited by campers and tourists by the millions each year and monitored on the ground and in aircraft by forest rangers. Alleged sightings, footprints, and fuzzy photographs and videos (many are faked) are what scientists call *soft evidence*. If we could find bones, that would be the hard evidence needed to convince the skeptics. Those who believe in Bigfoot say that bones don't survive. How often, they ask, do you ever see deer or bear bones? However, many bones of early hominids and other animals from Africa and elsewhere are found from various stages of evolution, millions of years old, but none from the manlike creature called Bigfoot. This may be because the animal doesn't exist.

Q. *What is the origin of the werewolf legend?*

A. It is anybody's guess, as these stories go back to the time of the ancient Greeks and perhaps earlier. The best-known legend is about a man named Lycaeon who sacrificed a child's flesh to the god Zeus. Apparently Zeus didn't like this and punished the man by turning him into a wolf. Based on this, the werewolf condition came to be known as *lycanthropy*. The word *werewolf* means "man-wolf" (from the Latin *vir* or the Gothic *wair*, both meaning "man"). The origin of the legend may also date back to times when Scandinavian warriors donned the skins of feared animals, such as wolves, before they attacked their enemies. Psychiatrists have treated people who imagine they can be transformed into a wolf under the influence of the full moon and then return to normal after a night of prowling for human flesh. The whole idea is known in many countries and under many guises. The true origin may never be known precisely.

Q. *Has any person ever been raised by wolves or other animals?*

A. It is really doubtful. The classic case, which is founded in myth, is that of Romulus and Remus, twins who were raised and nurtured by a she-wolf. Romulus and Remus were involved in the founding of Rome, and indeed Italian postage stamps portray the twins suckling from the she-wolf. Over the years there have been many stories of children raised in the wild by wolves, bears, or other animals. These children moved on all fours, ate raw meat, and could only make the sounds of the animals that raised them. A specific case was that of the wild boy of Aveyron, France, in 1797. Three hunters found him climbing trees and eating acorns and snarling like a beast. He was brought to Paris and exhibited in a cage, tragically enough. Most of such stories are without solid documentation. Those that seem authentic probably involve children who may have been abandoned or suffer from retardation or other mental deficiency. From the animal's point of view, we know of no wolf or other animal willing to—humanely—take a human child and raise it.

Q. *What is the difference between the legendary vampire and a ghoul?*

A. The lore of the ghoul grew from Arabic tradition and is much older than the vampire legend. The Arabs used the word *ghul* to describe a female demon that inhabited cemeteries and other lonely places. This ghoul lured travelers to their death and ate them. If attacked by a ghoul, it must be struck dead with one blow; otherwise, if struck twice, it would come back to life. Arabs used the tale to scare children into obedience and this may be the origin of the *Bogey Man*. Perhaps this is the reason that Western tradition depicts the ghoul as being fond of devouring small children.

The vampire is of Slavic origin, particularly the region of Transylvania, an area of Hungarian and Romanian heritage. Stories of vampires came from there during the eighteenth century. Perhaps its origins can be found in the few deranged persons who occasionally drank blood for any number of reasons. Blood symbolizes life and vitality. The idea was deeply implanted following the publication of Bram Stoker's book *Dracula*. Many books refer to Vlad the Impaler as the "model" for Dracula. Unlike the ghoul, the vampire is a bloodsucker whose victims turn into vampires. Most vampire lore was invented by moviemakers: sleeping in coffins, aversion to crosses and garlic, no image in a mirror, destroyed by driving a stake through the heart, and so on. The vampire bat, also lionized in film, is in actuality a pathetic little creature weighing about one ounce who laps up but does not suck blood. It can't lap up enough blood to kill anybody, but it can be a carrier of rabies.

Q. *Why was garlic thought to repel vampires?*

A. It isn't surprising that garlic was suggested because garlic cloves have been thought to cure just about everything except bad breath. It was said that if you carried around a piece of garlic, you'd never get sick. In addition, it was a powerful shield against evil spirits of various sorts and demons, including the vampire. Oddly enough, the plant was believed to have sprung forth from the footprint of the Devil when he was cast out of heaven. If so, the Devil landed in central Asia, where garlic seems to have originated. In the United States it is grown mostly in California. We recommend garlic as a flavoring but have serious doubts about its powers in most other respects. Considering its "fragrance," it is amusing that garlic is a member of the lily family.

Q. *Why are vampires supposed to cast no reflection in a mirror?*

A. This concept actually predates Dracula. In primitive times, one's reflection in a mirror, a quiet pool of water, or any shiny surface was thought to be one's soul or life force. In some societies, it is considered bad luck to see one's reflection in a pool of water for fear that some supernatural beast will steal it and thus take your life. Vampires were regarded as having no soul and therefore could not cast a reflection. The belief that bad luck ensues from breaking a mirror is similar, because if the mirror is broken while you are looking into it, you may lose your life. There is no scientific basis for these ideas.

Q. *Why is the Christian cross supposed to scare away evil beings such as vampires and devils?*

A. It may be that any cross is supposed to be a form of protection against evil because the cross as a symbol predates Christianity by many thousands of years. Crude crosses have been found on flat rocks and pebbles in ancient archaeological sites. In some cases, the meaning is uncertain, but scholars agree it is often a symbol of universality—the four arms of the cross stretching out in the four major directions of the compass to include everything in nature. With the coming of Christianity, the cross has come to more specifically stand for good versus evil. In occult and horror movies, movie producers have furthered this notion. One notable case where the idea of the cross as good has been corrupted is that of the swastika. The Nazis adopted this form of the cross in the mistaken idea it was a purely Germanic symbol. It has come to represent prejudice, death, and destruction.

Q. *Both the bat and the owl are creatures of the night. Why is the bat regarded as evil and the owl as a symbol of wisdom?*

A. Your assumption is not entirely correct. There are many cultures, perhaps the majority of cultures, which view the owl as an omen of evil and death equally with the bat. In other cultures, especially where the owl preys upon rodents and other animals that eat crops, the owl is thought of as good. Unfortunately for the owl, some cultures (such as the early Romans and even as late as nineteenth-century England) nailed an owl with wings outspread to the barn door to prevent hail and lightning from striking. The owl's reputation for wisdom seems apparent. With their wide-eyed stare, they look like they are thinking great thoughts. Not so the bat. The bat is indeed an ugly-looking creature whose association with the Devil and all things evil is firmly entrenched. Perhaps this is due in part to some varieties such as the vampire bat indulging in the drinking of blood and sometimes being a carrier of rabies. In contrast, some Oriental cultures consider the bat to be a symbol of long life. And in some places, the blood of the bat is thought to be an aphrodisiac.

Q. *Why is the Grim Reaper a symbol of death?*

A. The skeleton, perhaps since the earliest beginnings of humankind, has represented death. Some of the earliest drawings show a skeleton bearing darts with which to destroy those people whose time has come. Thus, death was personified generally in this way, with variations from culture to culture. During the Middle Ages, the idea of time was also personified as an old man with flowing mane carrying a long scythe. Humans long recognized the close connection between time and death, resulting in the combined symbol of time and death as the Grim Reaper, the skull peering out from beneath the cowl, carrying the scythe to cut the life stem of the living. Such representations sometimes carry an hourglass. A less grisly symbol is the Christian angel of death—less grisly, but just as final.

Q. *What is the "banshee" that wails as a warning of a person's impending death?*

A. It is a legend of Irish origin. Somehow the belief arose that a ghostlike creature associated with families of pure Irish blood would foretell with screeches and wails the death of a member of the family. Most legends have some basis in fact. In this case we suspect that a family pet such as a dog or cat did the howling coincident with the death of a family member. If the pet was out of sight, the conjuring of a supernormal being—the banshee—was not that unreasonable an explanation. The word seems to be derived from the Gaelic *bean-sith*, which means "fairy woman," a creature that acted as a guardian of certain aristocratic families. The Irish have other charming legends such as the leprechaun, a little old man with the power to disappear mysteriously and who guards a treasure of gold.

Q. *What is a zombie?*

A. Pure invention. The zombie is supposed to be a corpse bought back to life by means of voodoo. Belief in zombies is rife in Haiti but in few other places. The zombie is created by the use of poison drugs or enchantment, and used as a slave by the sorcerer. In this condition, the zombie has little power of speech and walks with a shuffling motion. If the zombie eats any salt, it is supposed to reawaken and may take vengeance against its master before expiring. The zombie legend likely derives from the fact that retarded people in Haiti behave like a zombie is supposed to, and superstitious people have accepted this belief.

Q. *How did the dragon myths originate?*

A. It is easy to imagine people long ago coming across giant skeletons of fossil dinosaurs exposed in rocky outcrops, many of which were indeed quite dragonlike, such as *Tyrannosaurus*. From there, the mythology of the dragon grew. Few cultures have no dragon legends. The beast is often depicted as being a scaly, snakelike creature with wings and sometimes two heads. It lives in deep caves guarding treasures, breathing fire and smoke when angered or disturbed. The dragon has become mostly a symbol of evil and may personify the Devil. Stories such as St. George slaying the dragon may represent the triumph of good over evil. However, in the Orient the dragon is regarded as generally a fine fellow and not evil at all, although when provoked, Chinese dragons have the power to eclipse the Sun. Dragons have a reputation for having excellent eyesight, and in fact the word is derived from a Greek word, *drakon*, meaning "sharp-sighted."

Q. *Is there any basis for the story about the Minotaur and the labyrinth?*

A. Like most myths and legends, it is probably a mixture of truth and fancy. There can be no biological basis for a creature half-man and half-bull devouring boys and girls. In the story, Theseus enters the labyrinth and slays the Minotaur and then finds his way out easily, having unrolled a string or thread as he progressed into the maze. Mazes or labyrinths were constructed in ancient times and seem to have originated in Egypt. They were associated with temples and rituals and may have been used as burial places. The ancient historian Herodotus described the Egyptian labyrinth at Arsinoe as a building containing three thousand chambers, half of them below ground level. However, there is no archaeological evidence of a labyrinth at Knossos in Crete where Theseus's exploits presumably took place.

Q. *What is a golem?*

A. In Jewish tradition, it would be similar to Frankenstein's monster: a creature brought to life from preexisting materials by means of magic. Fashioned from clay, the golem performed his master's bidding. On the golem's forehead was etched the word *emeth*, meaning "truth." The creature grew bigger each day, reaching great size. It was customary to erase the first *e* in *emeth* to alter its meaning to "he who is dead." Then the monster would collapse and be no more. If this was not done, the golem might do harm to its master. The broomsticks carrying water in *The Sorcerer's Apprentice* by Goethe, and set to music by Paul Dukas, immortalizes this legend. It also is a lesson in the misuse of power.

Q. *Why is the raven thought to be such an evil bird?*

A. It is an undeserved reputation, perhaps engendered by Edgar Allen Poe's "The Raven." As such, the raven is portrayed as a messenger of ill tidings and foreboding. In reality, there is no hardier and independent bird, which is also related to crows, but can support a wingspread of four feet! An omnivore, the raven can eat almost anything from dead fish to seeds. The raven's greatest enemies are humans, and yet it is not by law a protected species such as the eagle and condor. If a nestling or baby raven is raised in captivity, it can become a very good pet and be taught to say a few words, like a parrot. The bad reputation of the raven may be due mainly to its black color (although it may have some white feathers) that ally it with the night and things evil. It is amazing that a bird such as this, living on its wits, may live to be thirty years old, and some in captivity have lived to be nearly seventy.

16

GHOSTS, HAUNTED PLACES, AND LIFE BEYOND

Q. *Why do people believe in ghosts?*

A. It is difficult to pinpoint a universal reason why people would accept the idea that ghosts exist. As with most beliefs, it serves some sort of valuable social or psychological function. This is not to say ghosts do or do not exist. We cannot be sure. Since a ghost represents an aspect of a deceased person, belief in ghosts may provide reassurance to the living of survival after death. A more recent suggestion that moves closer to science involves the law of conservation of energy. This states that energy can be neither created nor destroyed. That being so, some have claimed that even after death the energy generated from the human body cannot be destroyed and can maintain a cohesiveness (as a ghost) for an indefinite time before dissipating. This is mere speculation, sounding scientific, but without evidence.

Q. *Is there any proof that there are ghosts capable of moving or throwing objects around in a "haunted" house?*

A. There is no solid scientific evidence that we are aware of. This type of ghost is known as a *poltergeist*, or throwing ghost. Many cases have been reported and investigated over the years. It seems most of these cases involve the presence of children. It is likely that many of the "phenomena" observed were simply pranks played by children on their elders for fun, or to attract attention to themselves. It is easy, when no one is looking, to throw an object across a room and then quickly play innocent. Yet some—even scientists—are gullible enough to accept these as genuine phenomena. Related to this is the idea of *telekinesis*, or production of motion in objects without physical means. Most data accumulated about this point to pure fakery. In one case, children were put in a room and told to try to bend spoons using only the power of their minds. When the investigators returned, many spoons were indeed bent. However, by closed-circuit TV, the children were seen to have stepped on the spoons. In this area of borderline science, it is best to be skeptical.

Q. *How can objects such as ships and planes be seen after their destruction?*

A. We suppose you might be thinking of the *Flying Dutchman*, one of the most famous ghost ships. It is true that there are numerous reports of mysterious ghost ships, planes, trains, automobiles, and other objects. Probably many if not all of these sightings can be attributed to misidentification, hallucination, mirages, and so on. Such phenomena are not amenable to scientific investigation. However, those cases, and there are several, that seem to repeat themselves might be critically studied. For example, it has been reported that a ghostly sailing ship enters the harbor of a town in Nova Scotia each December, and then burns and sinks before many witnesses. However, until instruments such as cameras are used to record the events, this report can't be evaluated. The idea of inanimate objects persisting similarly is highly speculative.

Q. *The Mary Celeste must be the greatest sea mystery of all. What do you think happened to the crew of this ghost ship?*

A. The *Mary Celeste* was found and boarded by the *Dei Gratia* sailing crewless near the Azores in December 1872. It might be the greatest sea mystery. The crew was never found. We will never know exactly what happened. But if we sweep aside all of the sensationalist ideas, certain facts emerge that permit a theory of what happened. We know that the single lifeboat was missing as well as the navigational instruments. A line attached to the vessel had broken. The cargo consisted of several hundred barrels of alcohol. If a ship is in trouble and the captain orders it to be abandoned, the first thing he would reach for are the navigational instruments. One theory is that the captain, Mr. Briggs, believed either that the flammable cargo of alcohol was about to explode or that the ship was about to sink, and ordered everyone into the single lifeboat, taking his instruments with him. Hoping to get back aboard, they kept a line tied from the lifeboat to the ship, but it broke, and they drifted away to perish, while the *Mary Celeste* sailed on alone to its meeting days later with the *Dei Gratia*. But who knows?

Q. *Do haunted houses actually exist?*

A. Throughout history there have been many thousands of reported instances of alleged haunted houses. The older the house, the more ghosts it seems to hold. This is why ancient castles always appear to be haunted. Intriguing as the idea is, it is a difficult area for science to address. There are many natural explanations for supposed hauntings. For example, stormy nights with strong winds will produce more creakings and noises, slamming doors and knockings, than usual. One house in England was famous because on stormy, rainy nights, a pool of "blood" appeared in an upper bedroom where a man had been murdered. It turned out that the "blood" was a pool of water stained red from leaves that had accumulated between the inner and outer walls. Rainwater trickling down through the walls took on a reddish color as it moved through the leaves and leaked into the room. Belief in haunted houses, and thus ghosts, gives us "evidence" of our immortality. Science cannot answer the question posed until one such house has been analyzed and the alleged phenomena scrutinized under strict scientific methods. So far this has not been done satisfactorily.

Q. *What would be the most famous haunted house?*

A. A top candidate would be Borley Rectory, which ghost-hunter Harry Price described as the most haunted house in England. He wrote two books about it. Borley Rectory was built in 1863 by the Reverend Henry Bull, and stood for seventy-seven years before being gutted by fire. Shortly after it was built, many reports of ghostly activity began to crop up. Supposedly there were mysterious noises and footsteps, a woman in white wandering around the garden, headless phantoms, and even a ghostly coach with horses galloping by. Even after it had burned, there were still reports of ghostly apparitions in the windows. Legends told of a fourteenth-century nun murdered on the site and human remains that were dug up in the basement.

About 1930, the Reverend Foyster and his wife, Marianne, moved into the house. Poltergeist activity supposedly increased with the tossing around of loose objects, doors slamming, and bells ringing. Harry Price investigated these phenomena, and even lived in the house himself for a year after the Foysters had left. During Price's residency, no ghostly occurrences were documented. He claimed that Marianne caused the poltergeist activity as a prank, and she allegedly confessed. At no time were rigid scientific tests conducted at the site, and since the rectory no longer exists, it would be difficult to resolve the controversy. However, it should be pointed out that Borley Rectory was infested with mice. They may have accounted for at least part of the "ghostly" noises.

Q. *I love to read Gothic novels. Why do they call them "Gothic"?*

A. During the twelfth century, a style of architecture was embodied in castles and other stonework of which the pointed arch, flying buttresses, and gargoyles were typical. Many such Gothic structures continued through the sixteenth century, and by the eighteenth century many were crumbling ruins or if intact, possessed an atmosphere of history dating back to the third century. This was when the original Goths, a barbarian race, invaded the Holy Roman Empire. With such a setting, authors weaved tales of terror often with overtones of the supernatural. If you wish to read what is perhaps the very first such Gothic novel, try Horace Walpole's *Castle of Otranto*, written in 1765. Also fair to add are Mary Shelley's *Frankenstein*, Bram Stoker's *Dracula*, and Mrs. Radcliffe's many novels. Stories of the Gothic type are as popular today as they were then.

Q. *Is there any truth to the reports that voices of the dead can be picked up on a tape recorder in allegedly haunted places?*

A. The method is to set up any ordinary tape recorder and allow it to record under conditions of absolute silence for five or ten minutes and then to replay the tape, listening carefully for ghostly whispers. Occasionally, sounds can indeed be heard that resemble a human voice whispering short statements such as "What's the matter" or "Missed you." The exact origin of these sounds is unknown, but some, perhaps all, are sounds produced by the internal workings of the tape recorder or subtle external noises not consciously recognized. These messages come very swiftly and softly and it can well be debated whether or not they are voices of the departed. It is one area of the unknown where you can investigate and judge for yourself.

Q. *Wouldn't the tape recording of the voices of ghosts in a haunted house be proof that ghosts exist?*

A. We would suggest being skeptical and look for a natural explanation if indeed some voices were recorded. The scientific approach is to consider alternative hypotheses and look for the most plausible explanation first. It could be a hoax; they happen every day. But let us say that a hoax can be ruled out. In one case, the ghostly voices turned out to be the faint feedback from a previously erased recording. In another case, twenty-six people were present at the recording session. Especially under the novel conditions of hunting for ghosts in a haunted house, it is difficult to keep that many people from doing a little whispering, perhaps not noticed by others, but picked up by the tape recorder with the volume turned way up. Unless rigorous conditions are maintained, little of scientific value can be learned. We have tried these experiments ourselves on several occasions without any positive results.

Q. *Is there any authentic account of ghosts haunting the Tower of London?*

A. The Tower of London was constructed many centuries ago and served as the scene of imprisonment and execution of many people, including several who were famous and illustrious. It is no wonder that legends of ghosts abound in this old fortress. Of course, many accounts have a natural explanation or are due to overactive imaginations. Repeated sightings have been reported of the ghost of Anne Boleyn, whom Henry VIII had beheaded in 1536. One encounter involved a sentry on guard duty in 1864 who saw a white figure come out of the darkness. When it refused to obey the command to halt, the sentry struck with bayonet, but it passed through the apparition and he fainted. In that condition he was found and accused of sleeping while on duty. At the court-martial, he claimed the "ghost" wore a bonnet but that the bonnet was empty. Other soldiers, as witnesses, corroborated his story and the court found him not guilty. Whatever it was, it seems something was going on in the Tower of London that night. Such as fellow soldiers in a mood for pranks.

Q. *Is there any truth to the story that at Fort Niagara, New York, there is a ghost of a headless soldier from Revolutionary War times who comes out of the water well at night?*

A. None. There is an underground conduit that intersects the well shaft and is big enough for a man to crawl through. According to officials at the fort, some soldiers played pranks on fellow soldiers on guard near the well by crawling to the well shaft and making mysterious groans and other noises. This would establish the idea of the ghost in the well. It would then be a short step to seeing someone walking near the well on a dark night and thinking it was a ghost. Why headless? The nightly figure was not recognized; therefore, it had no head! The story about the soldier is that there was a duel between two soldiers over a woman, and one cut off the other's head. To cover up his dastardly deed, he threw the body down the well and the head in the lake. People say the headless body comes up out of the well at night, looking for his head. Headless ghosts are always more scary than those that have one. We believe many ghost stories originate from pranks.

Q. *In the conversations some people claim to have had with the dead, how is the hereafter described?*

A. There has been a great deal of supernatural literature published that supposedly involves conversations with the dead. According to the literature, the "dead" speak of the "other side" in vague generalities and quasi-philosophical terms. When pinned down, we are told that the hereafter isn't much different than earthly life—there are houses, lawns, streets, and casual get-togethers for conversation. One dead person named Betty was asked by a medium if there were "electricity and oxygen and bricks and sticks and stones in her world." She answered yes, but that these things were dealt with "not in their obstructed aspects . . . but in their essence." This type of rhetoric answer is gibberish and sounds like a case of avoiding the real question. We suggest, until we know better, that such talks with the dead are manufactured. Those who are curious about the life beyond need only wait. They'll get there sooner or later.

Q. *Can the Ouija board really attract evil spirits?*

A. There are many who regard this as an amusing parlor game and others who take it quite seriously. The board itself contains all the letters of the alphabet, the numbers one through ten, and the words *yes* and *no*. Accompanying this is a pointer or planchette that can move around when one or more persons place their hands on it. Thus, words and numbers can be indicated, purporting to be messages from the beyond. We know of no documented case where such messages contained information not known to the participants. We suppose that on a dark, stormy night, those using the Ouija board might get carried away and unconsciously "help" the pointer to slide around and spell out a message that seems intelligible. We have made determined efforts with the Ouija board without success. The Ouija board gained great popularity some years ago when it was featured in a gripping ghost movie entitled *The Uninvited*. We recommend the movie and also a session with the Ouija board for fun, but do not expect to get any message from your dead Aunt Martha.

Q. *If the Ouija board has no psychic value, how can the case of Patience Worth be explained?*

A. The world-famous case of Patience Worth began about seventy-five years ago when Pearl Curran, a housewife living in the Midwest, began receiving communications via the Ouija board, starting with, "Many moons ago I lived. Again I come—Patience Worth my name." Gradually the Ouija message revealed the spirit of a girl who had lived in England during the seventeenth century. Over many years, Mrs. Curran served as the channel for a series of literary compositions. There were poems and novels of medieval and Victorian England, judged by experts to be of exceptional literary creativity, allegedly written by the disembodied spirit of Patience Worth. Investigators agreed that the quality and content of writing were far superior to the ordinary powers of Mrs. Curran. Was there really such a person, Patience

Worth, communicating from the spirit world? Or, as some scientists say, was it just an extraordinary case of secondary personality, that is, a creation of Mrs. Curran's unconscious mind?

Q. *Don't the "dancing coffins" on Barbados constitute proof of the supernatural?*

A. It is true that when sealed crypts at two burial sites were opened, the coffins had moved around, some of them even standing on end. Our instinct would be to look for a natural explanation first and think about the supernatural later, if necessary. One crypt had not been opened for a century, and a long-forgotten earthquake may have been the cause. Coffin movement at the Chase family vault is more difficult to explain. It had been opened more than once in recent years, and no earthquake was responsible. The Chase vault was carved from rock to a depth of four feet below ground. It is plausible that the coffins were shifted by the invasion and retreat of groundwater from below. The 600-pound coffins would displace more than 3,000 pounds of water and hence would float. The coffins were moved to another site on a hill and have not moved since.

Q. *I am fascinated with the idea of reincarnation. Does science offer any proof that we reincarnate?*

A. No. Many people around the world believe in reincarnation, often as a part of their religion. Its roots most likely lie in the fact that most people do not want to die. Central to the conviction, for many individuals, that reincarnation takes place is the experience called *déjà vu* ("already seen"). You come into a strange room and immediately feel you have been there before, but insist that you haven't. You might think you were there in a previous lifetime, but you may have been in the same room or a similar room during childhood and had forgotten. We do not offer this as a pat answer but only as one possible explanation. Several of those individuals claiming to be reincarnated stated they were queens, kings, princes, dukes, and other prominent personages in the former life. One consoling thought is the inheritance of characteristics. You, as an individual, may think, act, and resemble physically an ancestor who lived one thousand years ago.

Q. *Why do so many people believe they will live forever after death?*

A. Science has no answer as to whether or not this will occur after the death of the body. Perhaps so, and maybe not. As to why people believe this, we would note that a primary instinct in all creatures is that of survival. In lower animals, such as dogs and cats, this instinct is physical survival. But for humans, physical death as the final end seems unsatisfactory. It could be reasoned that people progressing through life often reach their end unrewarded for good works while other, ostensibly evil people escape without just punishment. Thus, there would have to be a hereafter in which these discrepancies were corrected if there was a just God. Most cultures believe in a continuance of life after death, and this could be a reflection of the instinct for personal survival and the desire to maintain connections with loved ones. In a sense, belief in immortality is the ultimate fulfillment of the biological instinct for survival.

Q. *Do most scientists believe in immortality?*

A. Many scientists adhere to formal religions in which the idea of immortality is a basic tenet. Other scientists who may not be so affiliated, as well as those who are, would no doubt subscribe to another form of immortality—that is, genetic. All organisms depend upon genes, the cellular elements containing all the information that defines their existence. Genes have the ability to produce exact replicas of themselves. The genes are, in a way, the essence—the germ plasm—of an individual. Through reproduction, this germ plasm remains alive as long as the descendants of the original individuals survive and reproduce. Those genes coding for characteristics that are successful in the evolutionary arena will remain, while those that are not will be weeded out. In a sense, then, if you want to be immortal, have plenty of offspring who grow up healthy enough to reproduce. Then there is the view that forever is too long, immortality a curse rather than a blessing, and death a release to return to the oblivion of preconception.

Q. *Don't "near-death" experiences prove immortality?*

A. Either humans continue on in some conscious form after death or they don't. No unequivocal answer is possible. We are suspicious of claims that a person, say, on an operating table died and came back to life and then reported that they saw a tunnel with light, spoke with long-dead relatives, and even saw God. Appealing as this may be, the temporary stoppage of vital signs can be viewed as *clinical* death rather than brain death. According to the conditions science imposes on itself, visions that a person experiences at the brink of death cannot be thought of as a brief visitation to the "other side." Another angle to this question is the fact that survival of the organism and its reproduction are the highest priorities of nature. That includes other creatures as well as man. For an imaginative and usually rational creature such as man, the concept of immortality is the ultimate way to survive, and is expressed in our cultural and religious systems as a wish fulfillment.

17

PROPHECY, SUPERSTITION, AND THE MIND

Q. *Can some people have genuine premonitions of future events?*

A. There are those who claim to have this power and also ordinary people who maintain they have had such flashes of some event that will happen in the near future. Usually the premonition is of something evil or unlucky. There is no scientific evidence that such ability can be exercised at will or even that it occurs at all. Consider that the more information you have bearing upon a future event, such as a presidential election, the better are the odds you will guess what will happen. If a person has a "premonition" that a close relative will die, that person may already have information about the relative's state of health and other circumstances that make the "premonition" probable. It should also be noted that only those premonitions that come true, or nearly so, are remembered and advertised. Premonitions that do not come to pass are forgotten and not recorded.

Q. *Is there any scientific basis for telling the future by reading a person's palms?*

A. People have been trying to do this for centuries. We know that palmistry was practiced more than 5,000 years ago and probably originated in either China or India. Modern science rejects the notion that the shape of the hand, its various lines and mounts, have any bearing upon one's destiny. Yet today people still spend money to have their palms read. A sharply observant fortune-teller can make almost uncanny statements about a person based upon the handshake and the physical condition of the hand. A hearty handshake indicates vigor and health, a limp handshake, the opposite; a calloused hand might be a manual laborer, a soft-skinned palm, an office worker. Palmists will also take into account other aspects such as the client's clothes, grooming, and vocabulary, to make statements about the education and finances of the client.

Q. *A palmist looked at my hand and said I would have three children. Should I believe this?*

A. If you did have three children, it might be concluded that the palmist had some magical ability or that your hand does reveal your destiny. Yet scientifically, such a conclusion is not necessarily correct. There is a certain logic in the idea that the hands indicate something about a person; one who gestures frequently, or bites nails may have a nervous disposition, for example. At the other extreme is the notion that specific lines on the hand reveal life expectancy, number of children, love affairs, and other aspects of life. Science does not support this belief or others that attempt to predict the future. One branch of palmistry—known as *onychomancy*—studies specifically the fingernails: their shape, thickness, coloration, unusual markings, and so on. One might use this information to conclude that the person, for example, has a fungus infection if the nail is a dead white color, but nothing is revealed, as science views it, about the person's fate. In the case of very dirty fingernails, something of one's character may be revealed. Finally, there is a chance of trickery. Some palmists have a client waiting room rigged for eavesdropping. Bits of personal information can be gleaned prior to the actual consultation and then revealed during the latter. It makes the palmist look very good indeed.

Q. *According to a numerologist, my name is a 2. What does this show about my character and prospects?*

A. Probably nothing. The pseudoscience of numerology has been around since the ancient Greeks, Romans, and Egyptians, and why not? Mathematics is the queen of the sciences and aids us in measurement of things and as symbolism for many ideas. However, some people believe that a number has a "vibratory" influence on individuals, and that words can be reduced to a number. Naturally, each number has a certain meaning. For example, your number 2 indicates contrasts and extremes. The number 7 represents the occult and the mysterious. There are several complicated systems, but a common one is to equate the alphabet to numbers as in $A = 1$, $B = 2$, $C = 3$, and so on. When $I = 9$ is reached, start over with $J = 1$, $K = 2$, and so on. If a person's name is Joe Smith, then we add $1 + 6 + 5 + 1 + 4 + 9 + 2 + 8 = 36$; $3 + 6 = 9$ and that is his name number. You can figure out your own the same way. The flaw in the notion that this number has great meaning is that it can change. A different number is perhaps found if we use Joseph Smith, Joseph T. Smith, Joey T. Smith, and so on. While it is an amusing parlor game, there is no scientific foundation on which its principles are based.

Q. *What is the idea behind crystal gazing to see into the future?*

A. The crystal ball and the rabbit materializing from a top hat are perhaps two of the most recognizable stage props in the whole field of illusion. Historically, the crystal ball has taken a variety of simpler forms. Civilizations as far back as the Assyrians, and later the Greeks and Romans, gazed into crystals, drops of water, polished metal, or gems. The concept was that one would see their reflected images or appearances that could be interpreted as mystic or religious experiences—perhaps prophecies of good or bad things to come. The function of the crystal seems to be to concentrate the gaze, but of course science does not attribute any importance to the crystal itself as the source of any visions. The so-called crystal ball is not a crystal at all, but rather a blob of clear, amorphous glass. If images are seen, they are more likely the product of suggestion, imagination, and that rich source of strange images, the human subconscious.

Q. *By using a biofeedback machine to increase alpha waves, can a person reduce stress and achieve a higher psychic state?*

A. We doubt it. The brain produces several types of waves, including alpha and beta waves. Generally, alpha waves are noted when a person is in a state of relaxation or dreamy reverie. Think of it in terms of cause and effect. Do the alpha waves generate relaxation or does relaxation produce alpha waves? Tests show that while one is attentive and has his eyes open, beta waves are produced. The simple act of closing the eyes generates alpha waves. There is no scientific proof that alpha waves create a higher "psychic" state. Science would judge that extravagant claims of purging bad habits and curing disease by means of biofeedback of alpha waves is charlatanism.

Q. *Can fortunes be told using dominoes?*

A. According to those who believe this, yes. The procedure is to place all the dominoes facedown, mix them, and then select three, one at a time. If the same one is drawn twice, its meaning is reinforced, and you are allowed to draw a fourth. In general, the drawing of a blank is bad, especially the double blank. Other doubles are considered good luck in various situations of love, marriage, job, and money. A double six is excellent for speculation in money matters such as the stock market. There is no scientific basis for any of this, nor for tea leaves or playing cards for that matter. Nonetheless, it is an amusing pastime if you don't take it too seriously.

Q. *Do tarot cards have any occult meaning?*

A. Our modern playing cards have probably been derived from them, but the tarot deck is much larger, consisting of seventy-eight cards with a variety of symbols covering all aspects of life—justice, death, fortune, family, romance, and so on. Tarot cards may have originated as early as the twelfth century and were presumably used for fortune-telling, but it is possible that the real origin of the tarot deck was as a tool or means of communication between tribes that could not speak each other's language. Thus, a deck of such cards would be very useful as symbols to interchange ideas. For telling fortunes, the deck would be laid out and the interrelationships of the cards permitted some interpretation—good, bad, or indifferent—that would be of interest, depending on the skill and imagination of the fortune-teller. There is no scientific evidence that tarot cards have ever predicted the future for anyone, and they should be regarded merely as a source of entertainment.

Q. *How can you say that the future is not seen in dreams when so many dreams are uncannily accurate?*

A. It is true that there are many seemingly prophetic dreams that psychics are quick to advertise, but we would have to dispute this on statistical grounds. Realize that all people sleep. Millions of people sleeping mean millions of dreams. If only one person in three had a dream they remembered each day, that would translate to 24 billion recallable dreams each year in the United States alone. Is it not likely that somewhere in this legion of dreams some of them by chance alone will hit upon a future event with what seems like, to many, uncanny accuracy? At the same time, we would point out that in scientific investigations, there is as yet no hard evidence for or against the idea that some dreams may be prophetic by virtue of some scientific law not yet recognized.

Q. *How many dreams does a person have each night and how long do they last?*

A. Dreams seem to take place during *REM* (rapid eye movement) periods. This is a light stage of sleep in which the eyes move and dart around under the eyelids. There are perhaps four or five REM periods a night which alternate with deeper stages of sleep. During a REM period the sleeper may have one or a series of dreams, some of which last only a few seconds while others are longer. The dreams we remember probably occur during the last

REM period of the night. In experiments, subjects deprived of REM sleep become extremely disturbed physically, and mentally irrational with hallucinations and blackouts. When allowed to sleep without interruption, these subjects experience an unusual degree of REM-type sleep. We might conclude from this that dreaming is important to maintaining good health.

Q. *If, in a dream of falling off a cliff, I do not wake before hitting bottom, could I be injured or killed by the fall?*

A. No, unless you had in fact gone to sleep on a cliff, had the same dream, and then fell. We have interviewed people who have hit bottom while falling from a cliff in a dream and they are still healthy. It is true that a particularly disturbing dream may cause restlessness during sleep, but it is not lethal. Dreams may represent our deeper fears and desires disguised as something else. A desire for a lot of money may take form in a dream of catching numerous fish. A fear of death may have a dreamer in a strange city, avoiding a particularly dark alley. A nightmare may occur with a loss of control by the subconscious, in which case a dread of something is portrayed in all its reality.

Q. *Concerning the biblical prophecy that the end of the world will come at the battle of Armageddon, is there such a place, and will the end of the world come in the near future?*

A. We need to distinguish here between, first, the destruction of our planet, and secondly, the end of the human race. In the latter case, we would assume that Earth would be habitable for whatever life survives humankind, perhaps beetles or ants. Life on our planet depends for its survival upon the continued functioning of the Sun. The Sun has enough light and heat remaining for many millions of years. Whether we humans behave ourselves sufficiently to take advantage of this planetary longevity is another matter. As for Armageddon, there is such a place in the Middle East in arid, sparsely populated country. It appears to be an unlikely place for armies to gather in a final struggle. With today's missile arsenals, any final battle is more likely to be planet-wide, consisting of distant combat, with more civilian than military casualties. Groups of people have gathered on hilltops at the end of centuries and millennia to witness the destruction of Earth, and it hasn't happened yet.

Q. *Where did the Gypsies get the reputation for having special abilities in fortune-telling and other occult matters?*

A. For centuries, Gypsies have been well known as wanderers and regarded by the local populace with suspicion, mistrust, and even fear. Perhaps the Gypsies took advantage of this and the gullibility of the curious to ascribe to themselves certain abilities with respect to things supernatural. After all, being nomads the Gypsies had a hard time making a living, and the telling of fortunes and selling of charms or amulets became just another way of making some money. The origins of the Gypsies are imperfectly known because they have little in the way of a written tradition. Their language, however, suggests that they migrated out of northern India around the fourteenth century. Today, they are found all over the world but their numbers are dwindling. Gypsies are true to their own codes of conduct, perhaps no better or worse than those of other societies. It might be noted that they have always suffered severe persecution. During World War II, Hitler wiped out thousands of Gypsies with the same zeal he used against the Jews. Some demographers estimate that Hitler killed 10 percent of the world population of Gypsies, mostly from eastern Europe. That they continue to endure is to their credit.

Q. *Did the prophecies of Nostradamus always turn out to be correct?*

A. There is considerable debate about this because Nostradamus's prophecies, made over four hundred years ago, were often vague and ambiguous. He wrote them as four-line stanzas, called quatrains, with a deliberate attempt to obscure their meaning. In many cases, the meaning was clear to some people after the event apparently took place. Not all students of Nostradamus agree on a specific interpretation. For example, the statement "The armies fought in the air for a long time" might suggest dogfights between enemy aircraft, or could simply mean a battle fought on a mountain. Likewise, the statement "Across the sky, a long running spark" could mean a meteor or a ballistic missile. The assemblage of quatrains has no chronology, that is, a quatrain may refer to the eighteenth century or the twentieth century. There are several books about Nostradamus. The interested reader might want to obtain one and make his own interpretations as to whether or not Nostradamus had the power to see into the future.

Q. *Was Mother Shipton as great a seer as Nostradamus?*

A. Mother Shipton, born in England in 1488, made several purported predictions, including the deaths of Cardinal Wolsey and Thomas Cromwell, Drake's defeat of the Spanish Armada, and the Great Fire of London. She supposedly wrote several prophetic poems that did not come to light until years after her death. Here is a small example:

> Carriages without horses shall go
> And accidents fill the world with woe.
> Around the world thoughts shall fly,
> In the twinkling of an eye...

This sounds like a forecast of automobiles and radio, but indeed it is a forgery by Charles Hindley, an English editor. Perhaps most of Shipton's "prophecies" were written by someone else after the event. Many of the quatrains of Nostradamus are equally suspect. Nostradamus is probably better known than Mother Shipton, but whether either could be called great is difficult to say because solid scientific proof that anyone can see into the far future is lacking.

Q. *Is it true that Hitler was a firm believer in astrology?*

A. Hitler was a believer in destiny and did consult with astrologers. To what extent this may have affected history is not exactly known. We know that Joseph Goebbels, Hitler's propaganda minister, dumped leaflets from planes over Allied territory with prophecies of the sixteenth-century astrologer Nostradamus that Germany would win the war. It was a good ploy, but it did not work because Nostradamus's prophecies could be interpreted in a number of ways. It is still believed in some quarters that Hitler stopped his Panzer divisions in France when they could easily have destroyed British troops at Dunkirk because an astrologer warned him to stop. We will never know for certain. It now appears from all the evidence that the so-called Hitler diaries are a fake. After Hitler wrote *Mein Kampf*, he didn't do much writing except by dictation.

Q. *Is it true that Mark Twain, born in a year in which Halley's comet appeared, predicted that his own death would occur on the occasion of its next appearance?*

A. Halley's comet returns to our solar system every seventy-six years and appeared in 1910 when Mark Twain (Samuel Langhorne Clemens) died. It was next seen in the sky during March and April of 1986. Mark Twain was of a psychical bent, and he claimed to have had a dream predicting his brother's death. It was Mark Twain who coined the term *mental telepathy*. That Twain did in fact "foretell" his own death with the reappearance of Halley's comet in 1910 is only speculation. In many of his writings, Twain shows himself to be a realist and a skeptic. Psychiatrists might infer that Twain believed it, and more or less made an autosuggestion that he would die at that time. It seems to us that there is nothing unusual about a person dying at age seventy-six.

Q. *Now that 1984 has passed, what can be said of George Orwell as a prophet?*

A. We are not so sure that it was Orwell's intent to predict what the technology of the 1980s would be like from his perspective of the 1940s. If so, he didn't do very well. His book *1984* makes little or no mention of computers, robots, arms races, spaceflight, or energy crises. The importance of oil is not addressed. Rather than technology, Orwell, who was an ardent anti-Communist, wanted to stress the evils of big government in restricting individual freedom. Oddly enough, there are smaller, less technically advanced countries around the world whose people enjoy less freedom than in many of the bigger, more advanced countries. We have heard that the original title of Orwell's book was *1948*. Because of constant editorial changes that had to be made (the book came out in 1949), they decided to place the setting farther into the future and therefore switched the last two numbers.

Q. *Do we have free will and control of our future, or is everything predestined?*

A. This question has been debated by philosophers for many years. The future can be thought of as "fluid" because a very minor factor or series of factors prior to an event can greatly alter what might happen to individuals or nations as well. For example, a cat runs across a road in front of a car and the driver puts on the brakes, slowing down the car and delaying his arrival at the next intersection by several seconds. If he had reached the intersection on schedule, he would have collided with a truck and possibly have been killed. Of course, it could be argued that a Supreme Being foreordained that the cat should run across the road. Similarly, if Adolf Hitler had never been born, World War II may never have happened. For individuals, it is a better idea to act now to determine your future (getting an education, saving money, and so on) rather than waiting around for the future to happen. True, there are things that no act of free will can accomplish—things that may be physically impossible, like being a lineman on a pro-football team if you weigh 110 pounds and are blind. Yet we see no compelling reason to assume that everything in life is predetermined and beyond our own control.

Q. *Does the daily horoscope in the newspaper have any real validity?*

A. Perhaps, if you are a devout believer in astrology. But even the astrologers are amused that people take it so seriously. An accurate horoscope, according to the astrologers, requires information on all planetary positions and not simply the Sun sign, which is what the daily horoscope is based upon. It seems to us that these horoscopes just dispense good advice such as "take time to relax" or "be careful of business investments." However, it seems absurd that, say, a Leo should avoid travel today, or Libras will find romance tonight. This translates to about 375 million people in the world not traveling today or expecting a romantic encounter tonight because the stars and planets are in a certain position. More scientific research should be done on the claims of astrology. Even if astrology were totally debunked, this doubtless would not change the minds of many people. Astrology has been around for 5,000 years and has always had its followers anxious to know their futures.

Q. *Doesn't the remarkable similarity between identical twins separated from birth show that astrology is not bunk?*

A. Only one out of every ten pairs of identical twins studied and reported on were totally separated at birth. Many were raised in the same kind of home environment and economic status. So, a vital question is, do persons of the same age and sex share strong similarities regardless of whether or not they are twins? Psychologist W. J. Wyatt and his colleagues checked this out by testing thirteen pairs of twins and twenty-five pairs of unrelated people of the same age and sex. Areas compared included jobs, politics, hobbies, favorite foods, and so on. The twins turned out to be more similar, but the unrelated pairs showed surprises. According to Wyatt, one pair of unrelated women were "both Baptist; volleyball and tennis are their favorite sports; their favorite subjects in school were English and math (and both listed shorthand as their least favorite); both are studying nursing; and both prefer vacations at historical places." Had these similarities been found in a pair of identical twins (who had been reared apart) they might have been used as evidence for astrology. We conclude the evidence is weak for any correlation between identical twins' similarities and astrological influence.

Q. *Did the ancient Romans practice astrology?*

A. At first they didn't. They seemed to have been more interested in worshiping many gods despite the fact that in lands and other peoples around the Roman Empire, astrology was hugely popular. In fact, the Romans were hostile to astrologers and in some cases persecuted and banished them. As we can see today, astrology is still popular and has persisted for at least 4,000 years. Thus, after a while, the practical Romans grudgingly gave in and allowed it to be practiced. It was an admission, as today, that people are entranced by the possibility that they might know their future fate—a desire that is very difficult to eradicate, despite the insistence of science that there is no foundation for astrology.

Q. *Is more crime committed during the full moon than at other times?*

A. Throughout history, many people have believed that the full moon exerts a strong influence on the human mind and emotions, even causing physical changes (as in the werewolf legend). Terms such as *lunatic* and *lunacy* are derived from the Latin word *luna*, meaning "moon." Two separate investigations of police records in Buffalo, New York, were made to check this out. Both studies were conducted over separate six-month periods. At the outset, even policemen believed that more rapes, murders, and other emotionally violent crimes occurred during the full moon. However, results showed no greater incidence of these crimes during the full moon than at other times. It would be interesting to compare similar data for other cities.

Q. *Why do some people have more luck than others?*

A. Like others, we have met individuals who do seem to be extraordinarily lucky in cards, lotteries, or "stumbling onto" the best job. Maybe that is why those of us not so well blessed with good luck resort to such helpful objects as a rabbit's foot or other charm. Others swear that certain numbers are lucky for them. We would have to say that chance plays a large role here. A person who wins a coin flip is not really lucky; there was a 50-50 chance of winning or losing and each new flip presents the same odds no matter how many times a person has won. According to some psychologists, the luck we have in life, good or bad, is brought about by our own previous actions and attitudes leading up to what may appear to be a lucky or unlucky event. The person who gets a very good job may have studied how to handle a job interview and taken the time to dress neatly. To those who did not do so, the person was "lucky." We would call it good planning. While life brings happy or tragic events on occasion which are beyond our power to circumvent, much luck can be controlled by ourselves.

Q. *How did the superstition of "knocking on wood" originate?*

A. Most superstitions are concerned with either bringing good luck or preventing bad luck. By knocking on wood we hope to prevent bad luck relating to something we express as a wish, such as "I think we'll win the game" or "I hope I get the job." It probably originated at a time when early peoples worshiped certain trees and thought there existed tree spirits who protected them. To touch the tree was to hold evil spirits at bay. To carry a piece of wood from a sacred tree was also a prudent idea. Archaeological evidence shows that the ancient Chaldeans, Egyptians, Persians, and others regarded trees as sacred. This is really not surprising when you think about the good that comes from trees—we eat the fruit, use the wood for building shelters and objects of art, and even cool ourselves in the shade of a tree.

Q. *Why is finding a four-leaf clover considered lucky?*

A. Clover in general has always been admired by man since the beginnings of agriculture and animal husbandry. Clover is excellent animal feed, rich in nutrition, and also a replenisher of nitrogen to the soil. The ordinary three-leaf clover may have been regarded as a symbol of the Trinity, as a further religious aspect. Rare things are usually highly regarded, and so the discovery of a four-leaf clover is an exceptional find. It has come to mean the bringing of good luck only if you do not give it away. It also suggests you will soon meet the person you will marry. In the eleventh century, the English minted a four-leaf clover coin. Each leaf could be broken off and used as a separate piece of money. Perhaps this is where the expression "rolling in clover" originated, to indicate wealth.

Q. *Where did the idea come from to make a wish with a wishbone?*

A. As we know, the one who gets the larger part of the wishbone will get the wish, or at least have better luck. Perhaps the convenience of its use for two people encouraged the custom, but we think there was a deeper meaning to it. Bone is the last part of the body to survive after death and so bones in general were believed by the ancients to have magical properties. They were protective. Even today in some societies, a bone is carried by those suffering from arthritis or other bone maladies. Fragments of the bones of saints are treated as holy relics by the Catholic Church. We still say, "I can feel it in my bones" when something is about to happen. Bone is made up of calcium, phosphorus, and carbonate minerals, which are interstructured with a fibrous protein called *collagen*. This gives the skeletal system both strength and elasticity. Unfortunately, bone becomes brittle with age and more susceptible to fracturing. About 99 percent of the calcium in the body resides in the bones. Only vertebrates (animals with backbones) have bones.

Q. *Where did the belief originate that a rainbow brings good luck?*

A. The association of a rainbow with good luck probably stems from the Bible. After the great flood, God told Noah He would never again send a flood to destroy the world and He set His mark or sign in the sky to symbolize this in the form of a rainbow. It is perhaps on this basis that some Christians think of people going to heaven by climbing the rainbow on a more worldly level to find a pot of gold at the end of the rainbow. However, when you examine early non-Christian cultures, you find that the rainbow often was a symbol of evil. A rainbow arced over a house forecast a death in that house; a man who reached the end of the rainbow would find death rather than gold; out of fear, children were brought inside when a rainbow appeared. So, various cultures have their own beliefs surrounding the rainbow. To the scientist, it is simply sunlight splitting into the component parts of light when it encounters droplets of water. By the way, moonlight can also produce a rainbow, although not as colorful. It is amusing to hear people speak of the "end of the rainbow" because the rainbow has no end. It is a circle. We see only the part visible above the horizon.

Q. *Why is the number seven considered so lucky?*

A. The roots of this belief seem to lie in astronomy. In ancient times, seven planets were known, and the Moon had phases that lasted seven days. Four of these phases equaled twenty-eight days, which seemed to govern the menstrual cycle, which in turn governed human life. The

Bible is abundant with references to seven, such as the days of creation and the seven marches around Jericho to bring the walls down. The number is still one of importance today: seven days in the week, seven wonders of the world, seven Deadly Sins as well as seven Christian sacraments, and many others, including the gambler trying to roll a seven. If your name has seven letters, that is supposed to make you rather special.

Q. *Why is a horseshoe thought to bring good luck?*

A. For many centuries the horseshoe has been regarded as not only a symbol of good luck but also as protection against witches or other forms of evil. Why? There are perhaps several reasons. The horseshoe is made of iron and forged with fire. Both iron and fire were viewed as magical by ancient peoples. Perhaps that is why the blacksmith was regarded as a special, almost supernatural figure. Even the horseshoe shape, resembling the crescent moon, had magical significance. Another aspect of the answer is the horse itself. Through the ages it was the horse that did a lot of our work—tilling the fields, carrying people and their goods, even being an instrument of war as cavalry, and in times of famine, a source of food. It is therefore no surprise that the horse and his shoes are held in special regard. Consider also that in more recent times in the nineteenth-century West, a man without a horse stood less chance of survival. That is why horse thieves were hanged.

Q. *Why is it thought that throwing salt over the left shoulder will avoid bad luck?*

A. Salt is a great preservative and the enemy of decay, as was known from ancient times. Also, it was thought to be allied to sterility, which is not a good thing. So in effect, salt was a mix of good and bad. We've all heard the expression "to rub salt in the wounds." It hurts. Centuries ago, it was administered as punishment to those who spilled and thus wasted salt. The latter were thought to be under the influence of the Devil. It became the practice for those who accidentally spilled salt to throw some with the right (good) hand over the left (evil) shoulder into the face of the Devil, as to avoid misfortune.

Q. How did the superstition arise that seven years' bad luck will result from breaking a mirror?

A. The mirror captures a person's image, and for ancient peoples, that image, whether reflected in a mirror or in a pool, was the soul of the person. A disturbance of the image was tantamount to loss of one's soul—death or disaster. With such bad luck in the offing, the question arose, How long would it last? The belief in ancient times was seven years because that was how long it was thought the body needed to replace itself and remove all of the bad luck visited upon the original body. Even today, some remote tribes object to anyone attempting to photograph them, fearing that the capturing of their image on film will cause harm to befall them.

Q. Why is it considered bad luck to walk under a ladder?

A. Ladders were made to allow people to climb, and in ancient times to get closer to the sky and thus spiritually closer to the sky gods. Archaeologists regard miniature ladders and depictions of ladders in ancient tombs as symbols of the ascent to a higher and better life. Consider that when a ladder is placed against a wall, it forms a triangle with the ground. To walk through this triangle is to destroy the good luck as well as to reject the symbolism of the ladder itself. In Christian times the triangle became a symbol of the Trinity, and the superstition persisted. Another more practical reason for not walking under a ladder is that you might get a bucket of paint dumped on your head.

Q. *Why do sailors consider the albatross a symbol of bad luck?*

A. Actually, they don't. For some sailors it is bad luck to kill an albatross. But for many sailors this is not so. Many sailors have gone "fishing" for albatross with line and hook, the bait skimming the surface of the water. In the past, sailors made tobacco pouches out of the skin of the bird, and pipe stems out of the hollow leg bones. There are thirteen species of albatross of which the wandering albatross is the most widely known. It is a truly remarkable creature with a wingspan of more than 11 feet and the ability to soar effortlessly in the wind for hundreds of miles. Studies of banded albatross have shown they can travel more than 300 miles a day. Essentially, they live their lives in the air, being grounded mostly during breeding times. Despite the fact that they are oceangoing birds, when taken aboard ship, they get seasick!

Q. *Is the appearance of a comet in the sky considered a good or bad omen?*

A. Through the ages it has been regarded as a bad omen, foreshadowing a famine or plague or disastrous war, even the end of Earth itself. The roots of this belief seem to be in the idea that if the normally orderly and predictable heavens are suddenly disrupted, a similar disruption would occur on Earth. Eclipses and meteor appearances can thus be placed in the same category. When such celestial events were indeed followed by some calamity, the notion of a cause-and-effect relationship was of course strengthened, although science would have claimed that it was coincidence. One's own perception is very important here. Julius Caesar was born at the time a comet appeared, and died at a similar time. Depending on what you thought of Caesar, the comet was either good or bad. We don't know of anyone who thinks Mark Twain brought misfortune on the human race; quite the contrary. Yet his birth and death occurred, as we know, at two successive appearances of Halley's comet.

Q. *Why do they smash a bottle of champagne against a ship when it is launched?*

A. It is a vestige of rituals performed in ancient times when a ship was built. It was believed that the sea and sky held many gods and goddesses that could either save or sink your ship. A libation, usually of blood, was given at the launching, and figureheads on ships often represented a protecting sea goddess. So the champagne bottle is a form of sacrifice. Early Greek ships had an altar aboard for sacrifice, and today officers salute the quarterdeck when boarding, perhaps not realizing that that was the place of the altar in early ships. Those who die at sea are buried at sea as quickly as possible because it is unlucky to have a corpse aboard ship. The captain traditionally goes down with his ship because the ship was supposed to be a living creature (female, of course) that would need a companion in "Davy Jones's Locker." There are many superstitions of the sea, of which these are only a few.

Q. *Why did they put figureheads on the bows of early sailing ships?*

A. Sailors were using these devices more than 3,000 years ago. A widespread practice was to paint a large pair of eyes on either side of the bow. Perhaps it was in the superstitious belief that the vessel would be better able to guide the seamen across dangerous waters. The use of these eyes seems to have originated with the ancient Egyptians, but the Phoenicians, Greeks, and Chinese painted eyes, or *oculi*, on their boats too. Later figureheads took the form of carved representations of animals or persons placed underneath the bowsprit and nine feet in length on larger ships. These figureheads stood for a particular religion or nation, so in a sense, figureheads were flags. They also served as battering rams during Roman times where the figurehead took the form of a unicorn or wild boar. In Colonial times, some British captains spent money out of their own pocket for elaborate and expensive figureheads. We speak of some people today with important titles but little authority as figureheads, much like those on the early ships—out in front leading but with no real function in the operation of the ship.

Q. *I've heard mention of the "rites of passage." It sounds mysterious. Just what are they?*

A. It really isn't mysterious at all, and it is most likely that the reader has participated in a rite of passage. Broadly speaking, rites of passage involve rituals associated with a change of social situation of a person or group of persons such as those passing from puberty to adulthood. It also includes marriage, childbirth, and death. There are ceremonies that accompany these events. Such rites have been practiced in primitive societies for many hundreds of years and involve things unfamiliar to us moderns such as sprinkling of blood on a newly married couple in Sarawak. Yet we do the same sort of things: first Holy Communion, throwing rice or birdseed at weddings, passing out cigars when a child is born, and many others. These customs were first described in detail by anthropologist A. van Gennep in his book, *Rites of Passage*.

Q. *When did men and women start to get married officially?*

A. Since the first groups or tribes of humans were organized thousands of years ago, there have been marriage ceremonies and rituals joining a couple together. From a survivability standpoint, such a rite insured protection of the young and cohesiveness of the group. Love had little to do with it. As societies became more complex, the reasons for marriage also became complex, involving economic and political considerations that persist even today. People still marry for security of the children, for money, or for sociopolitical purposes. In primitive times a person married within the tribe because there were perhaps no other tribes around. Even today, people tend to marry within their own class or religion (tribe). Long ago, they threw cereals at the married couple to insure fertility in a world short of manpower. Today we throw rice or ecologically correct birdseed. The bride is carried over the threshold, a symbol of when women were stolen from other tribes. Nowadays love is a motive for marriage in Western societies to insure ostensibly the permanence of the union. We expect marital unions sanctioned by society to always be around as long as human society exists.

Q. *Where did the custom originate of kissing someone under the mistletoe?*

A. In myth and legend, mistletoe is one of the most widespread and versatile among the objects of man's superstition and has been so for many centuries. It has been considered a good-luck charm, a talisman to ward off evil and the Devil, a cure for epilepsy, and a fertility symbol. It is probably for this latter reason that it came to be associated with kissing and marriage. In any case, a young man will find any excuse to kiss a pretty girl. Mistletoe is actually a parasite whose seeds are lodged on the bark of trees by birds. There it grows and feeds off the tree and does not grow in the ground. Perhaps this is another reason that early peoples perceived it as a magical plant. It is also thought to have been used by the mysterious Druids in their secret rites. A legend believed in some places states that the burning bush seen by Moses was actually mistletoe.

Q. *How old is the custom of the bride's "hope chest"?*

A. It is the relic of practices spanning many centuries. In earliest times, brides were purchased outright. The more attractive the bride, the higher the price. It was a straight business deal. With time, this evolved into the dowry which represented the money or goods a bride brought with her to a marriage as a sort of compensation for the bridal price paid in previous times. Thus, the marriage became something less crass than a simple business deal.

The hope chest seems to have evolved from the dowry. The father made the chest, and the future bride filled it over a period of years with items she made herself such as clothing or household goods that would be useful in the future marriage. Nowadays a hope chest can be purchased in a store and filled with items not much related to the person of the bride. A trousseau is similar in origin to the dowry. Most, if not all, accoutrements of modern marriage have ancient roots. Consider the bridal veil. It was worn originally as a sign of submission to the husband. Today's women would certainly reject that notion. Indeed, in some modern ceremonies it might be more appropriate for the groom to wear the veil.

Q. *Where did the idea originate that people who go to heaven play harps?*

A. The harp is a musical instrument of great antiquity. There is great probability that the association between the harp and heaven stems from a reference to it in Revelation 14:2. It is said that there were thousands of redeemed on Mount Zion "and I heard the voice of harpers harping with their harps." There is also a long-standing belief that the music of the harp wards off evil forces. There are Egyptian murals dating from 1200 B.C.E. showing large harps being played, but without any obvious religious connotation.

Q. *Where did the concept come from that angels have wings?*

A. We believe there are a couple of possibilities. It is recorded in the Bible that the angel Gabriel came "in swift flight" to Daniel and this suggests to some that the angel was flying. On the other hand, earlier peoples such as the Hittites and Assyrians depicted winged beasts on their monuments, and some think this may have influenced the Judeo-Christian concept of angels. Angels were conceived as messengers conveying information between God and man, and a fast messenger is more efficient than a slow messenger, so the power of flight would be important. While it is not a strict tenet of the Catholic Church, many believe that each person has a guardian angel. Angels are thought to be pure spirit, superior to man, and existing in large numbers. Some angels are described as rather tall—96 miles high, which is rather fantastic.

Q. *Do religions other than Christianity believe in angels?*

A. It is an essential tenet of Islam and is found in Jewish tradition. The popular conception of angels and their characteristics, however, is mostly of Christian origin. Angels were ranked in nine orders of which the archangels were one and certain angels had specialized duties as guardian angels, the Angel of Peace, and the Angel of Death. While angels are regarded as friendly and helpful to mankind, it was the fallen angel Lucifer who brought evil into the world. During the Middle Ages, churchmen did indeed argue over how many angels could dance on the head of a pin.

Q. *Why do drawings of saints show a yellow halo around their heads?*

A. This is the *aura*, thought by some to surround all people, but especially holy people with great spiritual power. Such representations are known from art dating from the fifth century and perhaps earlier. Sometimes the aura envelops the whole body and may be other colors such as green and violet. This is not strictly of Christian origin, as holy persons among the Moslems are depicted as being surrounded by flames rather than a glow. Also, even those thought to be evil (the Devil for example) are sometimes shown as being enveloped in a glow. In a real sense, each of us has an aura—one of heat—that surrounds us. That is how snakes can seek out and kill small mammals in the dark. While some people claim to be able to see this aura, there is no scientific evidence that this is possible.

Q. *Why do people in India make special trips to bathe in the water of the Ganges River?*

A. It is probably the most famous sacred river in the world as it is sacred to no less than 400 million Hindus. They believe the Ganges is the personification of the goddess Ganga, who came down from heaven. In addition, water itself is significant in Hindu religion especially as a means of purification. To bathe in the Ganges, especially at certain holy places on its banks, is to remove sin. Pilgrims return home with bottles of Ganges water. The dying often make a final pilgrimage to the Ganges for a last purification and may even die in the river; indeed, a few have drowned themselves in the river in the hopes of a better afterlife. A person is not allowed to spit into the river unless they spit on their hands first. Nor may women enter the river or cross it if they are menstruating. Although the 1,500-mile river is generally sluggish in its flow, during the 1876 flood season one million people were drowned in thirty minutes. The goddess Ganga seems to have her moods.

Q. *What is the explanation for some persons going into a kind of trance and speaking a foreign language they do not know in normal life?*

A. This is called *speaking in tongues*. In many of these instances, a person, either physically or mentally ill, will speak nothing but gibberish, but this is interpreted as an "unknown" or ancient language. Intriguing as this sounds, the so-called unknown language is never identified. We are reminded of an uneducated woman who suddenly started to speak Latin. It turned out that she had scrubbed floors and performed other menial tasks for several years in a monastery within earshot of clerics reciting Latin daily. In her feverish state, it is not surprising that she uttered a few incoherent words in Latin, but she most certainly did not recite the orations of Cicero in perfect Latin. We believe most such cases of speaking in tongues can be traced to a natural explanation and not a supernatural one.

Q. *There are a group of people—the Holy Ghost people of Appalachia—who handle poisonous snakes and drink strychnine. Why do they do this?*

A. To demonstrate faith: to "take up serpents" with the power of the Lord to protect them. In a region of high unemployment, poverty, and distance from business and industrial centers, the importance of religion and its social value is enhanced. Many groups meet three or four times a week for several hours at a time. At these meetings there is much praying and dancing with rhythmic percussion. This leads to a trancelike condition during which poisonous snakes are brought out and handled. Strychnine or other poisons are consumed, usually in small amounts. The snakes themselves are also in a torpor because of the percussion and are less dangerous. If bitten, most people in good health, whether in a religious trance or not, will recover. The body can tolerate strychnine in miniscule amounts, and indeed it has been used as an antidote for some cases of depressant-drug poisoning because it is a stimulant to the central nervous system, causing enhanced sensitivity to sight, sound, and touch.

Q. *When did Christmas tree decoration begin as a part of the Christmas celebration?*

A. The Christmas season originally was a pagan festival on the occasion of the winter solstice and the anticipation of spring. Indeed, celebration was condemned by the early Church and later Protestant groups. The ancient Scandinavians used yule logs and various evergreen decorations. The first use of a decorated tree seems to have occurred in northern Europe, specifically among the Germans during the sixteenth century or before. We know that German soldiers fighting on the side of the British in the American Revolution were already using Christmas trees. Yet the tree did not become a tradition in America until well into the nineteenth century and, in fact, the first Christmas card was not sent until 1844. It was about that time in England that, inspired by Charles Dickens's

Christmas Carol, the idea of helping the poor and hungry at Christmas became established. It is also worth noting that initially Christmas had nothing to do with children and presents. That was a later but happy development for younger people.

Q. *Is it possible that Christ was not really born on Christmas Day?*

A. There are some historical scholars who have argued that this is so. It is true that hard biographical information is very scanty. For such a world religious figure, Christ's life remains shadowy and unknown. Even the year of Christ's birth is uncertain. For a while, the year 4 B.C.E. was accepted. Later, the year 6 B.C.E. was thought more probable. December is not a likely month for the birth of Christ because shepherds do not tend their flocks outside during that time. The month of October has been strongly suggested. The Armenian Christians to this day do not celebrate the birth except on January 6. Christmas Day as the celebration of the birth of Christ was not accepted by the Church until the year 336. Christmastime coincides with traditional pagan festivals connected to the winter solstice which influenced the placement of Christmas. Even the Christmas tree, including the use of wreaths and garlands, is of pagan origin, having been used by the ancient Egyptians and other cultures many centuries before the birth of Christ.

Q. *How old was Mary when she conceived and gave birth to Christ?*

A. There is no historical or biblical evidence that we are aware of that addresses this question. Even in the New Testament, Mary is an obscure figure, mentioned only perhaps eighteen or nineteen times in unsubstantive terms. Women of that time typically married and bore their children at a young age, such as fifteen or sixteen, and there is no reason to imagine that Mary was any different. Scholars and theologians have spent more time on the virginity of Mary and the immaculate conception than they have on her historical reality. In comparing Mary with figures in other cultures, numerous virgin births were described, as in the case of the woman who ate a flower and became impregnated. Other stories indicate a conception by some deity, such as the conception of Helen of Troy by the god Zeus in the form of a swan.

Q. *Where did the three Wise Men who came with gifts for the baby Jesus come from and who were they?*

A. This is one of those history mysteries clouded with time and legend. They were otherwise known as the *Magi*, which we do know was a Persian cult of priests and magicians noted for their wisdom and knowledge of astrology. The names of the three Wise Men were *Gaspar*, *Melchior*, and *Balthasar*. Sometime following their deaths, their remains apparently were sent to Rome, and then in 1164 their bones were sent to Cologne, Germany, where they are interred in a high altar in the magnificent Cologne cathedral. We have no proof that these are the authentic remains of the three Wise Men. The trip to Bethlehem from Persia (now present-day Iran) would have been about 1,000 miles as the crow flies, and on camels making 25 miles a day would have taken them at least forty days. There is some speculation that the star of Bethlehem followed by the Wise Men was one that had flared into a nova.

Q. *How did Santa Claus become such a common symbol of Christmas?*

A. Toward the end of the fourth century in Asia Minor there was a very popular and beloved bishop named Nicholas who was said to work miracles and was particularly kind to children, and who, after his death, became the patron saint of schoolchildren. He became one of several gift-bearing figures that were associated with Christmas (like the three Wise Men). Dutch settlers in America called him *Sinter Claes* (which is "St. Nicholas" in Dutch) and the resemblance to Santa Claus becomes obvious. In 1823, Clement Clarke Moore wrote "The Night Before Christmas" and Santa Claus became entrenched in American tradition, but did not become popular in Britain until the 1880s. It should be noted that Santa Claus can't be everywhere. In Italy, it is the good fairy *Befana* that fills the stockings on Christmas Eve.

Q. *How did a rabbit bringing eggs get associated with Easter?*

A. The exact origin is obscure but the custom seems connected with old pagan rites of spring. The egg has always been a symbol of life and creation ever since early peoples watched birds hatch from eggs. Certainly the rabbit, or hare, with its prolific reproductive character is the common symbol of fertility. During Christian times, the Lenten fasting was particularly rigorous in some countries and often it was forbidden to eat eggs. Thus, many people might crave an egg as soon as Lent was over. Decoration of eggs for Easter originated at least 1,500 years ago. The most common Easter egg color has been red. A basis for this is one legend about a basket of eggs being placed at the foot of the cross during Christ's crucifixion, and blood dripping onto the eggs. Be that as it may, customs and ceremonies at Easter vary widely around the world.

Q. *I presume St. Valentine started Valentine's Day, but did he hand out candy hearts and heart-shaped cards?*

A. Surprisingly, St. Valentine had nothing to do with Valentine's Day. In fact, there were two Valentines, one a priest and the other a bishop, both martyred about the time the Emperor Claudius II was reigning in Rome. However, there was a pagan feast that used to take place on February 15 called *Lupercalia* in which the names of girls were placed in a large jar and the boys would draw names and pair off. After Christianity arose, they shifted this celebration to the fourteenth to coincide with the feast of St. Valentine and thus reduce the pagan overtones. By the way, people have been sending each other cards for thousands of years in one form or another. Some were sent as engravings on wood. One "card" was sent with the inscription on a single grain of rice! More than five billion cards are sent annually in the United States alone, including of course millions of valentines. The use of the heart and other devices is a modern fashion.

Q. *Do some people have a sixth sense of direction?*

A. We know how easy it is to get lost in the woods even with a compass, or in a strange town. If it seems that others have a "homing instinct" telling them which way to turn, it's not a sixth sense, experts say, but rather an aptitude for observation. Nomadic people or native pathfinders use their senses of sight, sound, and smell to discern direction. Add to this the mental map they construct. A similar "local reference" map is used by cab drivers in New York when they refer to "uptown," "crosstown," and so on. Even the blind have no sixth sense of direction but rather highly developed natural senses. There is growing belief that some animals, such as migratory birds, have traces of the mineral magnetite in the brain, serving as an aid in navigation within Earth's magnetic field.

Q. Have "psychics" located ancient archaeological sites?

A. Purported "psychic" or supernormal powers have indeed been applied to archaeological searches, although it is not considered a standard or practical technique in normal archaeological science. Psychic archaeology can be included with map dowsing and water witching as human potential shortcuts in the age-old quest for hidden treasures of one kind or another. Psychic experience in archaeology cannot yet offer positive proof of its value and is therefore not scientifically respectable. We usually read about someone with a strange ability to sense the presence of archaeological sites deep in the earth, or by holding an ancient artifact, perhaps an arrowhead in the hand. A person with proclaimed psychic talents is said to be able to "see" the maker of the artifact even across thousands of years. Scientists have their doubts.

Q. Are there people who can "see" with their fingers?

A. Like many strange mental phenomena, alleged eyeless vision or DOP (dermo-optical perception) has been reported for many years. The recent revival of interest in it was sparked by the case of a Russian girl who, it was claimed, could read print simply by moving her fingertips over the lines. Other cases of eyeless vision are stage acts performed by professional mentalists who claim special mental powers. Along with reading print, some who profess to have DOP are said to detect colors with their fingers, or sometimes even with their toes, elbows, or shoulders. Although entertainers and others who practice eyeless vision must be blindfolded, professional magicians who are well schooled in the art of deception know that blindfolds permit a tiny opening on each side of the nose. They suggest that the only true test of DOP is a metal box fitted over the head and under the chin to eliminate all vision. When subject to such rigorous testing, many of these gifted people lose their power, magically.

Q. *Can faith healers really cure disease?*

A. There are those who swear that ailments they had for years were healed spontaneously, and we will not argue with that. Yet we are not sure that faith healing necessarily transcends natural law in that it is miraculous. It is probable that many sicknesses were psychosomatic, that is, originating in the mind. This is not to say that such illnesses were not real—they may well have been. If a faith healer applies sufficient suggestion to the patient, who in turn sincerely believes in what is being done, this can effect changes in the mind which has strong control over the physiological processes of the body. Some faith healers are fakes who have confederates in the audience posing as sick or disabled persons. The healer and confederate, a *shill*, may even communicate by concealed radio. One may imagine dramatic "cures" resulting. Virtually all faith healers taint their work and motives by making a strong and aggressive appeal for money in the name of religion and live luxuriously as a result.

Q. *Is it a superstition that a ringing in the ears means somebody is talking about you?*

A. More often, we have heard that a *burning* of the ears signifies that somebody is talking about you. There is no scientific principle on which this belief could be based. A ringing in the ears is another matter. A surprising number of people have this annoying affliction known as *tinnitus* and it is well known to physicians. It may occur off and on, or constantly, and takes the form of a ringing, thumping, roar, or even the apparent sound of crickets chirping. There is no one cause of this malady, nor is there any specific cure. It can be caused by such diseases as syphilis, but also by tension or excessive use of tobacco and alcohol. Sedatives might be recommended. The ear was regarded in early societies as a place of entrance for evil demons and so people wore earrings as talismans against this. In addition, pierced ears were believed to aid those with poor eyesight.

EPILOGUE

If you have reached this point and read most of the questions and answers, you know it has been a long, varied, and hopefully interesting journey. Perhaps now we may reflect a bit. As you have seen, science probes into every nook and corner of our world, human curiosity its cutting edge. The most humble things up to the most grandiose are equally subject to scrutiny and investigation.

Think of something humble, such as a clump of soil in a field. You pick it up. It contains life—bacteria, perhaps even an earthworm. The soil consists of a range of particles of minerals from sand to minute bits of clay and silt, with dispersed decayed vegetation. Where did it come from? Maybe a rocky cliff on some distant shore, for it is decayed rock. What was it before then? Possibly a part of accumulating mud on an ancient sea bottom. During Earth's early molten stage, the matter in the soil may have flowed as lava down the slope of a volcano. And before Earth formed, where were these soil components? In the Sun itself. This humble clump of soil has also had a long and varied journey, as science reveals.

Our clump of earth represents only one of a countless number of the puzzles in the universe. The purpose of science is to sweep away the veil of mystery surrounding these puzzles, provide a clearer understanding of the real world, and lead to benefits through technology which increase the quality of life for all humans. It is an ongoing adventure.

There is a parallel here between science and those who think of themselves as humanists. Like science, humanists seek truth, using logic and reason. They respect tangible evidence and are skeptical of accepting conclusions on faith alone, or the assertions of others. They generally accept the idea that the world is here and now with immediate opportunity to achieve their personal potential, contribute to society, and help others. It is not a bad philosophy.

Finally, we will ask you a question rather than the other way around. Were you able to test your science IQ? We offered no rigorous little quizzes or exams simply because you are honestly the best judge of your own state of knowledge. We would say, however, that if you knew and understood something about the content of even half of our answers, you have done very well. If you enjoyed reading this book and it provokes you to explore other intriguing pathways of science, then this book has achieved its purpose. Do not forget that you, as a human being, possess a rational intellect, and as such you are also a scientist.

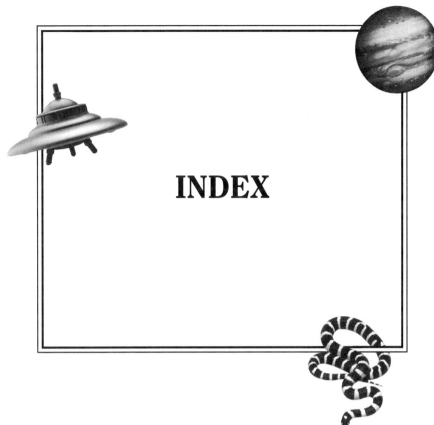

INDEX

abacus, 122
aborigines (Australia), 256–57
abracadabra, 277–78
acquired characteristics, 147, 174
afterlife, 251–52
   description of, 312
   ghosts and, 305
agriculture, 233
Akkadians, 239–40
albatross, 336
alchemy, 132
alien beings. *See* extraterrestrial beings
alligators, 163
aluminum, 80–81
amber, 77

anaerobic, 67
angels, 340–42
animals, 155–79. *See also* entries of specific animals
anticline, 91
antimatter, 20
Antony, Mark, 260–61
ants, 165–66
appetite, 211
archaeology
   excavations, 234
   frauds in, 229–30
   future, 236–37
   psychics and, 349
Archaeopteryx, 151, 177–78
Ark of the Covenant, 263

Armageddon, 58, 323–24
army ants, 165–66
asteroid, 48–50, 82
astrology
   compared to astronomy, 32
   Hitler and, 326
   horoscope, 328–29
   identical twins and, 329
   Romans and, 329–30
astronomy, compared to astrology, 32. *See also* outer space; solar system
aurora australis, 38–39
aurora borealis, 38–39
auscultation, 203
Australian aborigines, 256–57
australopithecus, 218, 219–20
avocado, 188

bacteria, 156
banshee, 302
Barnum, P. T., 175
barracuda, 160
basalt, 46
bat, 175–76
   symbol, 301
   vampire, 299–300
batteries, 134–35, 245
bauxite, 80
Beagle, 150
Bear Lake, 95
bears, 168, 170–71
bees, 176–77
belief. *See* religion
Bering Strait, 81
beverages, 133
Bible. *See also* names of specific biblical figures
   Armageddon and, 323–24
   creation and, 253
   life span and, 262
   rainbow and, 333
   science and, 139
big bang, 15, 20
Bigfoot, 296–97
biofeedback, 320–21
birds, 151, 177–79. *See also* entries of specific birds
Black Death, 206–207
black holes, 17, 18, 21
blimps, 129
blood
   leeching, 203–204
   spleen and, 213
   types, 204
   vampires and, 299
boa constrictor, 162
bogey man, 298
boiling point (water), 133–34
Boleyn, Anne, 311
bones, 332
Borley Rectory, 308–309
Boysen, Rudolph, 188
boysenberry, 188
Brady, Diamond Jim, 211–12
Braille, Louis, 211
brain size, 223
brainwaves, 320–21
bread, 233
Bronze Age, 218
Bubonic plague, 206–207
Buddha, 256
bufonenin, 190
bugs. *See* insects
Burbank, Luther, 187–88
buried alive, 213
butterflies, 72–73

cacao tree, 186
Caesar, Julius, 260–61
Cagliostro (magician), 278
calories, 134

camels, 172
carbonation, 133
Carlsbad Caverns, 110–11
Carnarvon, Lord, 257
Carter, Howard, 257
cashew tree, 184–85
Caspian Sea, 109
catacombs of Rome, 107–108
cats
   crib death and, 282–83
   domestication of, 235
   familiars, 282
   purring of, 174–75
   witchcraft and, 287–88
cattle mutilations, 289
caves
   Cro-Magnon and, 226–28
   formation of, 110–11
   prehistoric man and, 221–22
cedar tree, 182
cemeteries, 107–108
Centaurus X-3, 18
Ceres (asteroid), 49
chain reaction, 125
chameleon, 167
champagne, 337
Charon (Pluto's moon), 57
cheetah, 175
chemical weathering, 108
chicken, 155
chimpanzee, 148
Christian, Paul, 278
Christianity. *See also* Bible; religion; entries of biblical figures
   angels and, 341–42
   Romans and, 263–64
   spread of, 253
Christmas
   birth of Christ, 345
   Santa and, 346–47
   tree, 344–45

chromosomes, 144
civilization
   characteristics of, 231
   emergence of, 217–32
Claude, Georges, 126
clay, 108–109
Clemens, Samuel Langhorne, 326–27
Cleopatra, 260–61
climate
   deserts and, 115–16
   evolution and, 146
   global warming, 114–15
   ice age and, 114–15
   survival and, 221
cloning, 151–52
clover, 331–32
coal, 77
coffee, 269–70
Colorado River, 93
Colossus of Rhodes, 259
Columbus, Christopher, 40
comets, 50
compass, 248–49
computer, 122
continental drift, 87–91, 92
continents, 81–82
cork, 184
cosmetics, 136
cosmic rays, 36–37
cozmozoic theory, 41
craters, 47–50
   Earth, 49
   Moon and Mercury, 50–51
creationism, 146–47, 253
crime, 330
crocodiles, 163
Cro-Magnon man, 225–28
cross
   existence of, 264
   symbolism of, 259–60

vampires and, 300
crossing of fingers, 290–91
Crucifixion, 264
Crusades, 267
cryolite, 80
crystal gazing, 320
crystals, 64
   healing and, 209
   systems, 65–66
cubic zirconia, 78–79
curses, 277, 283–84
Cygnus X-1, 18

da Vinci, Leonardo, 131
dancing coffins of Barbadoes, 314
dark matter, 17
Darwin, Charles, 143, 149
dating, tree-ring, 130
David, 262
Dead Sea, 109–10
death, 301–302
degassing, 67
demons
   bats and, 176
   defined, 292–93
   familiars, 282
   possession by, 292. *See also* devil
dendrochronology, 130
dermo-optical perception, 349–50
deserts, 115
destiny, 327–28
devil. *See also* demons
   bats and, 301
   dragon myth and, 303
   familiars, 282
   Faust myth and, 291
   footprints of, 289
   garlic and, 299
   goat symbolism and, 286–87
   horned, 285–86
   jackal and, 290

musicical instruments and, 288
   origin, 285
   pitchfork and, 287
   salt and, 334–35
diet
   appetite, 212
   balanced, 193–94
   insects and, 200
   malnutrition, 197–98
   vegetarian, 196
   vitamins and minerals, 198
   weight loss, 195–96
dinosaurs
   crocodiles and, 163
   extinction of, 82, 83
   link to, 151
   migrations of, 81–82
directional sense, 348–49
disease, 205–208
DNA, 144
dodo, 178–79
dogs, 232, 234–35
doll, 280–81
domestication, 232, 234–35
dominoes, 321
DOP (dermo-optical perception), 349–50
Doppler effect, 22
dowsing, 111–12
Doyle, Arthur Conan, 284–85
Dracula, 299, 300
dragon, 303
dreams, 322–23
dynamite, 138

earrings, 351
ears, 350–51
Earth. *See also* earthquakes
   age of, 61
   atmosphere, 67–68
   continental drift, 87–91, 92

crust of, 92
end of, 58
evolution of, 153–54
Flat Earth Society, 117
formation of, 61
global warming, 114–15
hydrological cycle, 66
interior, 62
landscape, 71
origin of, 61
shorelines, 118
size of, 39
speed of, 43
surface processes, 102
tectonics, 88
trenches, 118–19
earthquakes. *See also* Earth
evidence of, 95
Hebgen Lake, 106
lake formation and, 95
magnitude of, 94
prediction of, 96
San Andreas Fault, 90, 93–94
San Francisco, 93
seismic waves of, 62
tectonics and, 88
thrust faulting, 91
tsunami and, 94
volcanoes and, 88–89
earthworms, 164–65
Easter, 347
Eckert, J. P., 122
eclipses, 39–40, 123–24
Edison, Thomas, 131
eels, 158–59
eggs, 347
Egypt (ancient), 243–47
curse, 257
mummification, 257–58
Einstein, Albert, 24, 131
electric eels, 158–59

electric shock, 208–209
elephants, 175. *See also* Mammoths
elevator, 129
ENIAC (computer), 122
entropy, 32
equilibrium, 33
Eratosthenes, 39
erosion
chemical (clay formation), 108–109
physical (sand formation), 117–18
ESP (extrasensory perception), 276
Euclid (mathematician), 121
Evelyn-White, H. E., 257
evil eye, 283
evolution. *See also* life
acquired characteristics 147, 174
animals and, 155–56
climate and, 145–46
creationism and, 146–47
defined, 143–44
earth and, 153–54
extinction and, 73–74
flight and, 150–51
genetic drift, 145
humans and, 217
to land, 70
excavation (archaeological), 234
expanding universe, 23
extinction
dinosaurs, 82–83
environmental change and, 145–46
evolution and, 73–74
extrasensory perception, 276
extraterrestrial beings, 26–27, 29–31
eyeless vision, 349–50
eyes, 201

Fahrenheit, 125
fairies, 284–285
fairy tales, 271
faith healers, 350
familiars, 282
farming , 233
fats, 194–95
Faust, Georgius, 291
finger crossing, 290–91
fish, 157, 159–60. *See also* entries of specific fish
Fizeau, Armand (astronomer), 124
flashlight, 134–35
flies, 222
flight, 150–51
floods, 104, 253–54, 333
Flying Dutchman (ghost ship), 306
food
   appetite, 212
   groups, 193–94
   preservation of, 199
Fort Niagara, 311–12
fossils, 72, 73
   fuels, 77
   spirifer, 73
   sunlight, 77
   trilobite, 72
Fountain of Youth, 134
foxes, 174
Franklin, Benjamin, 126–27, 132
frauds, 229–30
frigatebird, 178
fruit flies, 177
fruits, 188–89

galaxy, 16, 23, 24
Galileo, 37, 44, 56
gall bladders, 173
Ganges River, 342
Garden of Eden, 254–55
garlic, 299

gasoline, 137
gems, 78–79
gene pool shift, 146
General Sherman tree, 182
genes, 144–45, 316
genetic drift, 145
genius, 131–32
ghosts, 305–14
ghoul, 298–99
giants, 262
Giffard, Henri, 129–30
giraffe, 173–74
glaciers, 113–14
global warming, 114–15
goats, 286–87
Goebbels, Joseph, 326
Goethe, 291
golem, 304
Goliath, 262
Goodyear Blimp, 129
gorilla, 148
Gothic, 309
gradation, 102
grail, 265
Grand Canyon, 93
grapes, 190
graveyards, 268
Great Pyramid, 244–45
Great Sphinx, 246
Great Wall of China, 244
Greek fire, 138
Greenland ice cap, 115
Grim Reaper, 301–302
Grimm, Jacob and Wilhelm, 271
gunpowder, 138–39
guyot, 120
Gypsies, 324

Hades, 293–94
Hadrian's Wall, 248
Hahnemann, Samuel (physician),

210
hair, 212–13, 214
Hall Process, 80
Hall, Charles, 80–81
Halley's comet, 326–27
Halloween, 293
halo, 342
hanging valley, 104
hare, 171
harps, 340–41
haunted houses, 307–10
headless horseman, 311–12
healing
　crystals, 209
　faith, 350
　Lourdes, 266
　magnets, 209
health and nutrition, 193–214
Hebgen Lake, 106
Helen of Troy, 242
hell, 294
Henry VIII (king of England), 311
herring, 160
hibernation, 170–71
hiccups, 201
Hindenburg disaster, 130
Hitler, Adolf, 196, 326
Holocene, 218
Holy Ghost people (Appalachia), 343–44
Holy Grail, 265
homeopathy, 210–11
Homer, 241
hominids, 217
Homo erectus, 225–27
Homo habilis, 225–26
Homo sapiens sapiens, 231
hope chest, 340
horns, 286
horoscope, 328
horseshoe, 334

Horus, 202
hot springs, 97–98
Hubble, Edwin, 23
human beings
　earliest, 217
　evolution of, 217
　frozen remains of, 85
　longevity and, 198–99, 262
　population and, 152–53
　preservation of, 85–86
　warfare and, 228
hurricane, 116
hydrological cycle, 66
hyena, 170

ice. *See also* polar ice; glaciers
　Greenland ice cap, 115
　ice age, 114–15
　slipperiness of, 128
ice age, 114–15
identical twins, 329
imitative magic, 280
immortality, 315–16
ink, 135
insects. *See also* entries of specific insects
　flight of, 150–51
　immunity of, 150
　intelligence of, 169
　preservation of, 78
　training of, 169
intelligence
　brain size and, 222–23
　of chimpanzees and gorillas, 148
　in evolution, 222
　of foxes, 174
　of insects, 169
iridology, 201
iron, 240
Iron Age, 218

jackals, 290
Jefferson, Thomas, 189
Jews
　golem and, 304
　Rome and, 260
Joan of Arc, 266
joiner, 27
Jupiter moons, 44, 55

Keely, John, 124
kerosene, 137
Kilauea volcano, 97
kissing, 339–40
"knock on wood," 331
koala bear, 168
Krakatoa volcano, 98
krypton, 68
kwashiorkor, 197

labyrinth, 303–304
lactovegetarians, 196
ladders, 335
Lake Baikal, 109
Lake Superior, 109
lakes, 109–10
　Dead Sea, 109–10
　earthquakes and, 95
　size of, 109
Lamarck, Jean-Baptiste, 147
landslides, 106
languages, 237–39
Laplace, Pierre Simon, 18
law of use and disuse, 147, 174
Leaky, Dr. Mary, 220
leaning tower of Pisa, 106–107
leeches, 203–204
lemon tree, 185–86
lie detector test, 270–71
life, 42, 71. *See also* evolution
life-forms. *See* extraterrestrial beings

lightning, 126–28
lion, 162
lion fish, 158
Lippershey, Hans, 44
lithification, 247
Loch Ness monster, 295–96
logography, 238
Lourdes, France, 266
luck, 330–36
"Lucy," 219, 220
lycanthropy, 297

Mach, Ernst, 128
Magi, 346
magic, 276–84
magma, 65
main sequence, 17
malnutrition, 197–98
Malthus, Thomas, 152–53
mammoth
　compared to mastodon, 86
　preservation of, 85
　resource, 84
man. *See* human beings
mare, 46
Marlowe, Christopher, 291
marriage
　hope chest, 340
　origin, 339
　threshold, 268–69
　wedding cakes, 267–68
Mars, 52–55
　canals, 54
　life on, 52
　volcanoes, 54
Mary, 345–46
Mary Celeste (ship), 307
Masada (fortress), 260
mastodon, compared to mammoth, 86
mathematics, 121, 245

Mauchly, J. W., 122
measurement, 123
  of speed of light, 123
  temperature, 125–26
Medici, Cardinal, 125
Mendel's Law, 144
mental telepathy, 327
Mercury (planet), 50–51
mercury, 135
metallurgy, 240
Meteor Crater (Arizona), 82
meteorite, 41
  life's origin and, 41
  meteoroid, 41
  Tunguska (Siberia) explosion and, 42
meteors, 41
Methuselah, 261–62
Milky Way, 16, 18, 20, 36
mineral water, 133
minerals, 64–65
mining, 241
Minotaur, 303
mirrors, 135, 300, 335
mistletoe, 339–40
monkey, 147
monomineralic, 64
monsters, 295–304
moon
  formation of craters, 47
  full moon, 330
  Jupiter, 44, 55
  man in the Moon, 46
  origin of, 45–46
moraines, 113
Moses, 92–93, 253, 261
mosquitoes, 163, 222
Mother Shipton, 325
Mt. Saint Helens volcano, 98
mummification, 247, 257–59
mummy, 258

mushrooms, 190–91
music, 340–41
myrrh, 84

natural gas formation, 75–76
natural selection, 143
navigation, 248–49, 348–49
Neanderthals
  caves and, 111
  classification, 219
  fate of, 224–25
  intelligence, 223–24
  origin of, 224–25
Neolithic Age, 218
neon, 126
Neptune, 57
Nero (Roman emperor), 270
Newberry, Percy Edward, 257
Niagara, Fort, 311–12
Nightingale, Florence, 204–205
1984 (Orwell), 327
Noah, 104–105, 140, 253
Nobel, Alfred, 139
northern lights, 38
Nostrodamus, 325–26
number seven, 333–34, 335
numerology, 319
nutrition and health, 193–214

oak tree, 184
Occam's razor, 69
occultations, 40
oceans
  effects of sea level, 83, 84, 113–14
  evolution and, 70
  origin of, 66
  polar ice and, 84
  pollution and, 163–64
  seafloor spreading, 89
  seafloor topography, 120
  tsunami, 94, 119

octopus, 162
oil
  boiling in oil, 270
  depth of, 118–19
  drilling of, 76
  formation of, 75–76, 77
  kerosene and, 137
  pollution and, 164
  production of, 76
Olduvai Gorge, 217
onychomancy, 319
opal, 79–80
opalescence, 79
orange tree, 187
Origin of Species (Darwin), 150
Orwell, George, 327
ostriches, 179
Otis, Elisha Graves, 129
Ouija board, 312–14
outer space, 15–33. *See also* solar system
  asteroids, 48–50
  astrology/astronomy compared, 32
  big bang, 15, 20
  big crunch, 16
  black holes, 17, 18, 21
  double-star system, 25
  life-forms, 26–31
  neutron star, 21
  nova, 17
  planets, 25–27, 61–62
  pulsar, 21
  quasar, 19, 24
  radio transmission from, 28–29
  star formation, 16–17
  white dwarf star, 21
owls, 301

pagan, 260, 286
Paleolithic Age, 218
palmistry, 318–19
pandas, 168–69
paper, 137
papyrus, 137
paranormal, 275
Parris, Samuel, 281
particle physics, 37
passenger pigeon, 179
pegmatites, 65, 241
pemmican, 199
peneplain, 102
perpetual motion machine, 124
Petrie, Flinders, 257
Petrified Forest (Arizona), 112–13
philosopher's stone, 132
Phoenicians, 242
physician's symbol, 202
phytoplankton, 75
pictographs, 238
pig, 207
Piltdown forgery, 229–30
pimples, 201
piranha fish, 157–58
Pisa, Italy, 106–107
Pisano, Bonanno, 107
pitchfork, 287
planets, 25, 26–27. *See also* entries of specific planets
  alignment of, 43
  formation of, 25, 61–62
plankton, 156–57, 164
plants, 67, 181–92. *See also* entries of specific plants
plate tectonics, 88
Pleistocene Age, 218
Pluto, 57
polar ice, 84
pollution, 163–64
poltergeist, 306
polygraph, 270–71
Ponce de Leon, 134

population explosion, 152–53
pork, 207
potato, 191
pottery, 235–36
praying mantis, 169
precognition, 277
predictions, 325–26
premonition, 317
prominences, 38
prophecy, 325–27
psychokinesis, 276
pulsar, 21
pyramids, 244

Qantas Airways, 168
quasar, 19, 24
quicksand, 105–106

rabbit, 171, 347
radio telescopes, 19
radio transmission, 28–29
rainbow, 333
rainforest destruction, 183
Ramsey, William, 126
rapid eye movement (REM), 322–23
raven, 304
Reaumur, Rene-Antoine, 125
Red Sea, 92–93, 261
red shift, 22, 23
redwood, 182
refrigeration, 199–200
reincarnation, 314–15
relativity theory, 24
religion, 251
   afterlife, 252, 256–57
   ancestor worship, 252–53, 255–56
   ancient, 252
   Buddhism, 256
   creationism, 253
   cross, 259–60
   Crusades, 267
   evolution and, 146–47
   halos and, 342
   immortality, 315–16
   monotheism, 252
   pagan, 260
   polytheism, 252
   rainbow, 333
   reincarnation and, 314–15
   science and, 139
REM (rapid eye movement), 322–23
Rhodes (Greek island), 259
ring of fire, 90
ringing in ears, 350
rites of passage, 338
rivers
   Nile, 101
   pollution and, 163–64
   process, 102
   resources of, 101–102
rock
   categories, 63
   clay formation and, 108
   striations of, 113–14
Roemer, Olaus (astronomer), 123
Roman catacombs, 107–108
Rome (ancient)
   Christianity and, 263–64
   Egypt and, 260–61
   Jews and, 260

saints, 342
Salem witchcraft trials, 281–82
salt, 334
San Andreas Fault, 90, 93–94
sand
   formation of, 117–18
   quicksand, 105–106
San Francisco, 93
Santa Claus, 346–47

Sargasso Sea, 103
Sargon (ruler), 239–40
Saturn rings, 56
Schliemann, Heinrich, 242
scientific method, 68
seahorse, 159
seas. *See* oceans
secondary recovery, 76
seeds, 182–83
sense of direction, 348–49
Sequoia, 182
serpent handling, 343–44
SETI, 28–29
seven (lucky number), 333–34, 335
sharks, 161
sherds, 236
ships
    champagne and, 337
    figureheads and, 337–38
    Flying Dutchman, 306
    Mary Celeste, 307
Shipton, Mother, 325
shock, 212–13
shorelines, 118
Shroud of Turin, 264
Siberia
    Lake Baikal, 109
    mammoths in, 84–86
    Tunguska explosion, 42
snakes, 162, 166–67, 343–44
snowflakes, 65
soap, 136
solar flares, 37
solar system, 35–58. *See also* outer space; sun; entries under specific planets
    asteroids, 48–49, 50
    comet, 50
    cosmic rays, 36–37
    Earth's movement and, 43
    eclipses, 39–40

Galileo, 44–45
meteorites, 41–42
meteors, 41, 47
northern lights, 38–39
origin, 35–36
planetary alignment, 43–44
space programs, 47–48
sunspots, 37–38
telescope, 44–45
Tunguska explosion, 42
sonic boom, 128
sound speed, 128
space programs, 47–48
speaking in tongues, 343
specialization, 74
speed of light, 123
speed of sound, 128
sphinx, 246
spiders, 162–63, 166
spirifers, 73
splenectomy, 213
splenomegaly, 213
squid, 162
St. Valentine, 348
stars, 16–18. *See also* solar system; sun
    double-star system, 25, 40, 82
    formation, 16–17
    neutron star, 21
    nova, 17
    white dwarf star, 21
steady-state theory, 16
steam, 133–34
stethoscope, 203
stingrays, 158
Stone Age
    Australian aborigines, 256–57
    defined, 218
stress, 212–13
striations, 114
strychnine, 343–44

subduction, 89
sun
  death of, 58
  double star, 82
  effects of, 38
  northern lights and, 38
  origin of, 35–36
  sunspots and, 37–38
sun stones, 249
supernatural, 275
superstitions, 330–38, 350–51

tarot cards, 321–22
tea, 269–70
tectonics, 88
telekinesis, 306
telescope, 44
Thera (volcanic island), 93, 99, 261
thermometer, 125
thunder, 126–28
thrust faulting, 91
tigers, 162
time machine, 130–31
tinnitus, 351
toadstools, 190–91
tomato, 189
tools, 149, 218, 220–21
torture, 270, 279
totem, 256
totem poles, 255
Tower of London, 311
tower of Pisa, leaning, 106–107
Travers, Morris, 126
tree-ring dating, 130
trees, 182–87, 331, 344–45. *See also*
  names of specific trees
trenches, 118–19
trichinosis, 207
trilobites, 72
Trojan, 241–42
Troy, 241–42

tsunami, 94, 119
tuberculosis, 205–206
Tunguska (Siberia) explosion, 42
Twain, Mark, 326–27
twins, 329
Typhoid Mary, 207–208

UFOs, 30
UNIVAC (computer), 122
Uranus, 57

Valentine's Day, 348
vampire, 299–300
vampire bat, 299
vegetable, 188–89
ventricular fibrillation, 208
Venus
  phases of, 44
  surface, 51–52
Venus's-flytrap, 181, 191
Vikings, 248–49
violet shift, 23
vitamins, 198
volcanoes
  age of, 98
  depth of, 97
  earthquakes and, 88–89
  formation of, 62
  hot springs and, 97–98
  Krakatoa, 98
  Mars and, 54
  Mount St. Helens, 98
  Red Sea parting of, 93, 261
  tectonics and, 88
  Thera, 93, 99, 261
  tsunami and, 98–99, 261
voodoo, 302

warfare, 228
Washington, George, 204
water, 133–34

waterfalls, 103
watermelon, 189–90
water witching, 111
weathering
   chemical (clay formation), 108–109
   physical (sand formation), 117–18
werewolf legend, 297
Werner, Abraham, 253–54
whale, 156–57
whirlpools, 103
winds
   deserts and, 116
   hurricanes, 116
   seed distribution and, 183

wine, 190
wings, 341
Wise Men, 346
wishbone, 332
witchcraft, 278–82
wolves, 172, 297–98
wood, 331
worms, 164–65, 203–204
Worth, Patience, 313–14
written languages, 237–39
wrought iron, 240

zebra, 162
zircon, 78–79
zombie, 302